Great Thinkers, Great Theorems

William Dunham, Ph.D.

THE
GREAT
COURSES™

PUBLISHED BY:

THE GREAT COURSES
Corporate Headquarters
4840 Westfields Boulevard, Suite 500
Chantilly, Virginia 20151-2299
Phone: 1-800-832-2412
Fax: 703-378-3819
www.thegreatcourses.com

William Dunham, Ph.D.

Truman Koehler Professor of Mathematics
Muhlenberg College

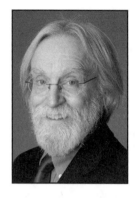

Professor William Dunham is the Truman Koehler Professor of Mathematics at Muhlenberg College. His undergraduate degree is from the University of Pittsburgh (1969), where he earned membership in Phi Beta Kappa during his junior year and received the M. M. Culver Award as Pitt's outstanding mathematics major in his senior year. From there, he went to The Ohio State University as a University Fellow from 1969 to 1974. At Ohio State, he finished his M.S. in 1970 and his Ph.D. in 1974, with a dissertation in general topology written under Professor Norman Levine.

Professor Dunham has taught at Hanover College in Indiana as well as at Muhlenberg College. He has received teaching awards from both institutions as well as the Award for Distinguished College or University Teaching from the Eastern Pennsylvania and Delaware Section of the Mathematical Association of America. In addition, he has twice been a visiting professor: first at Ohio State from 1987 to 1989; then, in the fall of 2008, at Harvard University, where he was invited to teach an undergraduate course on the work of the great Swiss mathematician Leonhard Euler.

In 1983, Professor Dunham received a summer grant from the Lilly Endowment to develop a "great theorems" course on the history of mathematics. This led not only to the class itself but to his first book, *Journey through Genius: The Great Theorems of Mathematics* (John Wiley and Sons, 1990), which became a Book-of-the-Month Club selection and has since been translated into Spanish, Italian, Japanese, Korean, and Chinese. Another spin-off was a series of summer seminars funded by the National Endowment for the Humanities (NEH) and directed by Professor Dunham at Ohio State in 1988, 1990, 1992, 1994, and 1996. As mathematics seminars, these were something of a departure from the usual NEH fare, but Professor Dunham's idea of portraying great theorems as works of (mathematical) art carried the day.

In the wake of that first book came two more in the 1990s—*The Mathematical Universe: An Alphabetical Journey through the Great Proofs, Problems, and Personalities* (John Wiley and Sons, 1994) and *Euler: The Master of Us All* (Mathematical Association of America, 1999). In the present millennium, Professor Dunham wrote a fourth book, *The Calculus Gallery: Masterpieces from Newton to Lebesgue* (Princeton University Press, 2005), and edited a fifth, *The Genius of Euler: Reflections on His Life and Work* (Mathematical Association of America, 2007). These books garnered various honors: *The Mathematical Universe* was designated by the Association of American Publishers as the outstanding mathematics book of 1994; *Euler: The Master of Us All* received the Mathematical Association of America's Beckenbach Book Prize in 2008; and both that volume and *The Calculus Gallery* were listed by *Choice* magazine among the outstanding academic titles of their respective years.

In addition to these books, Professor Dunham has written a number of articles on mathematics and its history. Among these are papers that received the George Pólya Award in 1993, Trevor Evans Awards in 1997 and 2008, and the Lester R. Ford Award in 2006. These awards, presented by the Mathematical Association of America, recognize excellence in mathematical exposition. In addition, Professor Dunham has provided mathematical journals with a cartoon ("Math Prodigy Field Trip") and a poem ("For Whom Nobel Tolls"), although these are unlikely to challenge the reputations of Charles Schultz or Emily Dickinson.

Over the years, Professor Dunham has presented numerous talks on mathematics and its history. These include lectures to students and faculty at scores of U.S. colleges and universities, ranging from Amherst to Bowdoin to Carleton, from Davidson to Denison to Dickinson. He has also addressed the scientific staff at businesses (Texas Instruments, Air Products) and governmental agencies (Goddard Space Flight Center, the National Institute of Standards and Technology), and he has performed on a larger stage with appearances on the BBC, on National Public Radio's *Talk of the Nation* "Science Friday," and at the Smithsonian Institution. Perhaps his most unusual venue was the Swiss Embassy in Washington DC, where Professor Dunham gave a 2007 lecture on Euler, a son of Switzerland whose tercentenary was being celebrated.

Professor Dunham is pleased to add to his resume this course for The Great Courses. As he has done throughout his career, he is happy to share the genius of great mathematicians, and the beauty of their great theorems, with a new audience. ■

Table of Contents

Table of Contents

Table of Contents

Great Thinkers, Great Theorems

Scope:

In this course, we meet some of history's foremost mathematicians and examine the discoveries that made them famous. The result is something of an adventure story. As with all such stories, the characters are interesting, if a bit eccentric. But rather than leading us to unexplored corners of the physical world, these adventurers will take us on a journey into the mathematical imagination.

Everyone is aware of the utility of mathematics. Its practical applications run from engineering to business, from astronomy to medicine. Indeed, modern life would be impossible without applied mathematics. No human pursuit is more useful. But the subject has another, more aesthetic side. It was the 20th-century philosopher Bertrand Russell who described mathematics as possessing "… not only truth but supreme beauty—a beauty cold and austere, like that of sculpture, without appeal to any part of our weaker nature, without the gorgeous trappings of painting or music, yet sublimely pure, and capable of a stern perfection such as only the greatest art can show." It is *this* vision of mathematics as an unmatched creative enterprise that guides us as we explore the genius of great thinkers and the beauty of their great theorems.

We begin with mathematicians of ancient times. Chief among these are Euclid and Archimedes, whose specialty was geometry and whose achievements, even after two millennia, remain as fresh as ever. It was Euclid, of course, who gave us the *Elements*, the most successful, influential mathematics text of all time. His successor, Archimedes, stands as the most creative mathematician of the classical era for determining, among other things, the area of a circle and the volume of a sphere. Additionally, we meet Thales, Pythagoras, and Heron—three other Greek geometers who left deep footprints. Together, these individuals should give a sense of the rich mathematical tradition of that distant era.

With the fall of Rome, the world's intellectual center shifted eastward; thus, we next consider the justly esteemed Muhammad Mūsa ibn al-Khwārizmī, who in the 9th century, solved quadratic equations by (literally) completing the square. Then we follow the trail back to Renaissance Italy with the colorful Gerolamo Cardano, who published the algebraic solution of the cubic equation in 1545. This brings us to Europe in the 1600s, sometimes called the "heroic century" of mathematics. It was a time that saw the appearance of logarithms, number theory, probability, analytic geometry, and by century's end, the calculus. A list of individuals responsible for this explosion of knowledge reads like a scholar's hall of fame—Fermat, Descartes, Pascal, Newton, and Leibniz—each of whom we shall get to know.

Building on these achievements, the irascible Bernoulli brothers and the incomparable Leonhard Euler pushed the mathematical envelope throughout the 18th century. And the discoveries kept coming in the 19th, with the work of Gauss, Cauchy, Weierstrass, and Germain. We end with two lectures on Georg Cantor, who did battle with the mathematical infinite in what proved to be a shocking departure from all that had come before.

Besides meeting these remarkable characters, we should come to appreciate the masterpieces they left behind. These are the "great theorems," which are to mathematics what the "great paintings" are to art or the "great novels" are to literature. We shall consider some of these in full mathematical detail, among them, Euclid's proof of the infinitude of primes, Archimedes's determination of the area of a circle, Cardano's solution of the cubic, Newton's generalized binomial expansion, Euler's resolution of the Basel problem, and Cantor's theory of the infinite. An understanding of such works will reveal the extraordinary level of creativity that is required to produce a mathematical masterpiece.

This mathematical/historical/biographical journey will carry us through the centuries and across frontiers of the imagination. In the end, our course should provide an appreciation of the artistry of mathematics, a subject that Bertrand Russell aptly characterized as the place "…where true thought can dwell as in its natural home." ∎

Theorems as Masterpieces
Lecture 1

It is a wonderful thing that progress in mathematics does not come at the expense of the past. In that sense, math differs from so many other subjects, so many other fields.

Rather than a course that teaches mathematical skills, this lecture series is a journey through the history of mathematics, focusing on the foremost mathematicians of all time—the great thinkers—and the extraordinary masterpieces they produced—the great theorems. We will regard theorems as the products of the creative imagination of mathematicians, and we will judge them by certain characteristics: elegance, or economy, and an element of unexpectedness, or surprise.

In mathematics, we will find that great theorems are not generally superseded by new discoveries or advances in the field. If a theorem is proved once, it is proved forever, and we will see results from ancient Greece that were still being used in 18th- and 19th-century mathematics and are cited today. In math, we do not discard older ideas; we build ever upward.

If a theorem is proved once, it is proved forever.

There are two general lines of attack for proving a theorem: direct and indirect proof. With direct proof, we reason directly from a hypothesis to a conclusion. Indirect proof might also be called proof by contradiction. With this strategy, we assume that the hypothesis is true, but the conclusion is false. From that assumption, we reason our way to a logical contradiction. If we reach such a contradiction, we may presume that the conclusion we were trying to prove is true. When a proof is complete, we end it with "Q.E.D.," the abbreviation for the Latin *quod erat demonstrandum* ("which was demonstrated").

One of the logical issues we will explore in this course is the contrast between a statement and its converse. Consider the statement: If *A*, then *B*. The converse reverses the role of the hypothesis and the conclusion: If *B*, then *A*. In mathematics, if we prove a statement, we almost always flip it around and ask: What about the converse? It is not always the case, however, that if a theorem is true, its converse is also true. As we will see in a future lecture, the converse of the Pythagorean theorem is not true in all cases.

We will begin the course with a lecture on pre-Euclidean mathematics before turning to three lectures on Euclid himself and his *Elements*, one of the greatest mathematics textbooks ever written. From there, we will move to Archimedes and his formula for finding the area of a circle. Then, we will jump to medieval Islam, to the world of Muhammad ibn Mūsā al-Khwārizmī, an Arabic scholar who wrote a well-known treatise on algebra. Next, we will meet Gerolamo Cardano, a strange character from 16th-century Italy who published the first proof of the solution of the cubic equation. In the 17th century, we will meet Isaac Newton and Gottfried Leibniz, co-creators of the calculus. Toward the end of the course, we will encounter Leonhard Euler, the most prolific mathematician in history, and his successor Carl Friedrich Gauss. We will conclude with Georg Cantor in the 19th century, who gave us the theory of the infinite, a profound, radical, and exciting idea.

As we begin our journey in the world of mathematics, Bertrand Russell's characterization seems particularly apt: "Remote from human passions, remote even from the pitiful facts of nature, the generations have gradually created an ordered cosmos where pure thought can dwell as in its natural home." ■

Suggested Reading

Hardy, *A Mathematician's Apology.*

Questions to Consider

1. A valid theorem about whole numbers is: "If *m* and *n* are even, then *m* + *n* is even." State the converse and determine whether or not it is valid.

2. Prove the following theorem: "If the perimeter of a triangle is 35 feet, then at least one of its sides must be longer than 11.6 feet." HINT: Do this by contradiction; that is, begin by assuming that the conclusion is false.

Theorems as Masterpieces
Lecture 1—Transcript

Hello. My name is William Dunham. I am the Truman Koehler Professor of Mathematics at Muhlenberg College. I would like to welcome you to our course about great thinkers and great theorems.

This course will be a journey through the history of mathematics where I will focus on the foremost mathematicians of all time, our great thinkers, and the extraordinary masterpieces that they produced, our great theorems.

Math courses come in various shapes and sizes. You have probably taken many in your day. Some of them are designed to impart mathematical skills. You might have a course in Algebra to learn the skill of solving equations. A course in Trigonometry would give you the skill of working with right triangles. This is not a course about acquiring skills.

Other math courses are about applications. Nothing has greater applications than mathematics. There are great courses in statistics for instance. You learn statistics and you can apply it to election results, to drug testing, all sorts of things. Differential equations, the more sophisticated application to explain where the satellites will be next week or whether a bridge will stand up. These are wonderful courses. This course is not about applications either.

If you want an analogy, if you want a parallel, this course is most like a course on art history. You can easily imagine a course. In fact, let me give a parallel title: *Great Artists, Great Paintings*. A course on that would introduce you to the great artists of history. You would meet Vermeer, Rembrandt, Monet, and Van Gogh. You would place them in their historical context, figure out what was going on in the art world before they arrived, what they did, what they passed on to the future artists. You would also look at their work. You would have to see their masterpieces. You would not be able to appreciate these artists without seeing what they did, to see a Vermeer interior, a Rembrandt self-portrait, a Monet water lily, a Van Gogh starry night. You could have a course titled *Great Authors, Great Novels*, a literature course where you would meet Austin, Dickens, and Twain and read their works.

In this course, we are going to want to see the masterpieces as well as just talk about the great creators. So this will be a course where understanding the theorems is essential. Now I should point out it is not an encyclopedic survey of the history of mathematics. That is impossible any more than you could study every painter or every author. I will ask you to trust me in my selection of the mathematicians and their results. I think I have a good lineup for you.

In this course we will regard theorems as the products of the creative imagination of our mathematicians. The theorems will be artworks. Like artworks, we can judge them by certain characteristics. One of the most obvious is just the incredible genius that emerges that can pop off the page, that can pop off the screen, as you can see these results unfurl.

There are other maybe more narrow qualities that you are looking for in a great theorem as a great work of art. One of them is elegance, economy. Some of these theorems are extremely elegant where they go from premise to conclusion in what seems like the quickest possible way. You are convinced that it could not be improved upon no matter what you did. This is the quintessential proof of that result. When you see that, you know you are in the presence of a mathematical beauty.

Another feature of a beautiful theorem is a certain element of unexpectedness, an element of surprise. It is not just in the conclusion as to what it is that is being proved. We will see examples where the conclusion is surprising indeed. Sometimes it is in the manner in which the mathematician reaches it. You are following along with the argument and you do not know where this is headed. Where is this proof going, and then suddenly, poof, a surprise occurs, and there it is. The result falls out. When you see that, that is a special, special moment.

These lectures will take various forms. Some of them will be quite biographical as we look at the great thinkers. You'll want to meet some of these people. You'll want to meet Archimedes, learn about his life and his memorable death. Cardano, the most bizarre figure from the history of mathematics. Newton, Leibniz, Euler, Gauss, these are people you should get to know. I have a colleague who tells me he became a historian because it allowed him to make such interesting friends. I think you will meet some interesting friends here.

Other lectures, maybe the typical lecture, will be sort of a hybrid. There will be some biography, there will be some history, and there will be some math. The math might be fairly simple or might even be just presented in an expository rather than a technical fashion. Then there will be some lectures that are entirely mathematical. The whole lecture might be devoted to deriving one result, to proving one theorem. This is where we are going to look carefully at these great landmarks.

These will be challenging. A whole lecture devoted to one theorem, I understand that. Here is my pledge that I will try to break these arguments down into little bite-sized pieces so you can follow each step, and at the end step back and say yes, I see it. I get it. That is my goal, to make these understandable. In spite of the challenge, if you can get it, it is extremely rewarding to be able to follow one of these arguments. Why is that? Well, there is the self-satisfaction you get, to be able to see what one of these great thinkers did. It is like when you understand a poem. You get it; you see what the poet was up to. That makes you feel good about yourself and that is worthy.

The other reason this is a good undertaking is because you will get to appreciate genius, really appreciate genius, by looking at these arguments as they unravel. Wait until you see some of the genius that we will encounter. It is all over the course. Euler, the 18th Century mathematician who is probably my favorite, has proofs that are just spectacular. I am going to show you two of these.

When I teach this, sometimes students will say, "How did he think of this? How did Euler come up with this?" You can look at the argument and you can make some suggestions, well maybe he was led from this step to that step by certain ideas or certain something that he saw there. In the final analysis, I end up saying, I do not know. He was Euler. He was a genius. You just cannot explain it. It is like asking: Why did Shakespeare put the balcony scene in *Romeo and Juliet?* Well, it is a great scene. You can see it is spectacular. What made him think of it? Well, he was Shakespeare. This is what genius looks like. By following these theorems, you will get to see what genius looks like.

I have drawn the analogy between this course and courses on the history of art and literature. Let me say in one way this course differs from other kind of surveys. In particular, imagine a course on the history of medicine, *Great Physicians, Great Treatments*. There might be such a course. It would be fascinating, and you could go back to the ancient times and study the physicians, Hypocrites and those folks, and then how they treated ailments. What you would do however as you came up to the modern period is discard those old treatments. In the old days, they thought your disease was caused by some imbalance in bile and phlegm and all of these bodily fluids. They would treat you by bleeding you or something. We do not do that anymore. We have advanced beyond those ancient medical practices.

In mathematics however, that does not happen. If a theorem is proved once, it is proved forever. We are going to see results from ancient Greece that are still being used in 18th to 19th Century mathematics. We will sight them in this course. They were proved. It is a wonderful thing that progress in mathematics does not come at the expense of the past. In that sense, math differs from so many other subjects, so many other fields.

The mathematician Hermann Hankel put it best perhaps when he said this, "In most sciences, one generation tears down what another has built, and what one has established another undoes. In mathematics alone each generation adds a new story to the old structure." In math, you are not discarding anything, you are just building upwards. If you like a more succinct version of that, Oliver Heaviside said, "Logic can be patient because it is eternal." We are going to be looking at eternal theorems in this course.

One thing I want to do in this introductory lecture is sort of limber up the mathematical muscles. I want to talk about two points of logic that we will be seeing over and over again as the course proceeds. One of them is: How do you prove a theorem? What is the strategy? I am going to mention two lines of attack when proving a theorem, and give examples.

One of them is direct proof. It seems like a pretty obvious sort of thing. The idea there is you reason from the hypothesis to the conclusion. The theorem will always have that form; if this happens, then that happens. In a direct proof, you just reason from the hypothesis to the conclusion directly, as the name suggests.

Let me give you an example. I will draw this example from the realm of number theory. The number theory is dealing with the whole numbers. The simplest system 1, 2, 3, 4, ... sometimes called the natural numbers, the counting numbers. My theorem here that I am going to prove directly is coming from that arena.

Here it is, theorem: If the whole number M is a perfect square, then so is $M + M + M + M$. That is the statement of the theorem. Now, what is a perfect square? A perfect square is like nine, 3×3; 16, 4×4. So what I am saying is that if M is a perfect square, and you add it to itself four times, you end up with a perfect square.

Proof, here comes a direct proof of that. First of all, let us say because M is a perfect square, we know that $M = k^2$ for some whole number k. I am not getting specific here. If M is 16, it is 4^2, 4×4. If M is 25, it is 5×5. I do not know what M is; it is just a perfect square. I do know it looks like $k \times k$ for some whole number k.

Now my challenge is to look at $M + M + M + M$ and see if that indeed comes out to be a perfect square as well. You could check this with 9. M is 9, is a perfect square, $9 + 9 + 9 + 9$ is 36, 6^2. You could check it with 25. You could check it with 49, but you cannot check it for every number. There are infinitely many perfect squares. I cannot hope to prove this by just listing the cases. There are too many. I am proving it this way, in a more abstract fashion.

Let's go back to my proof now. We had $M = k^2$ for some whole number k. Now I look at $M + M + M + M$. Well, that is $k^2 + k^2 + k^2 + k^2$. I see there I have four k^2s. So, $M + M + M + M = 4k^2$. That is a perfect square because $4k^2 = (2k)^2$. $2k \times 2k = 4k^2$. So indeed, $M + M + M + M$ is the square of $2k$ if k is a whole number, $2k$ is a whole number, voila, $M + M + M + M$ is a perfect square. The proof is over and we write Q.E.D. Now, what is that? That is an abbreviation for the Latin *quod erat demonstrandum*. It means, the proof is over. You always like to see that coming up, Q.E.D. So that was a direct proof.

There is another strategy though for proving a result. It is called the indirect proof or the proof by contradiction. This is more subtle. It is very cleaver, a nice bit of logic going on here. Here is what you do. You still want to prove that the hypothesis implies the conclusion. For an indirect proof, you assume the hypothesis is true, but assume the conclusion is false. From that, reason your way to a logical contradiction to a flaw. Get yourself into logical hot water. If that happens, you may conclude that the conclusion you were trying to prove is true. That by negating it, by rejecting that conclusion, if you get to a contradiction, you have shown in a sense that the conclusion cannot be false. It has to be true and you've proved your result indirectly.

We will see this in the hands of some great mathematicians. Archimedes does a wonderful proof for the area of a circle in which he proves it with double contradiction. Wait until you see that one. Other mathematicians will invoke and employ this slick method of proof. Let me give an example just so you see one here. This one will come from the domain of Algebra, simple Algebra. Suppose this is my theorem: There is no solution to the equation $x^2 + 2x + 1 / x + 2 = x$. There is my equation. The theorem says there is no solution to that. There cannot be. Well, again, you can see I cannot check this. I could let $x = 5$ and stick it in there and it does not solve it, so 5 is not the solution. I could check 7. That is not the solution. I cannot check every number.

If I want to show there is no solution, I have to be a little more logically sophisticated. Here is what I will do. My proof will be by contradiction. That will begin by my assuming exactly the opposite of what I am trying to show. I am trying to show there is no solution; okay, suppose there is one. I am going to assume there is a solution to that equation. I have to call it something, suppose it is a. What that means is, if I substitute a into that equation, it is going to work. If I replace x by a, I will get that $a^2 + 2a + 1 / a + 2 = a$. That would follow from the assumption that there is a solution.

Is there a contradiction there? Well, you might not see it, it takes a little work, but there is. Here is how I will extract it. Let me cross multiply that fraction, so you leave the $a^2 + 2a + 1$ on the left, but on the right you take the $a \times$ that denominator $a + 2$. So now we would know that $a^2 + 2a + 1 = a(a + 2)$. I can multiply out on the right side, distribute the a across, so that $(a \times a) + (a \times 2)$ and you would be left with this fact, $a^2 + 2a + 1 = a^2 + 2a$. Now, subtract a^2 from each side, gone. Subtract $2a$ from each side, gone. What remains on the left is 1, and what remains on the right is 0. So, you would have concluded that $1 = 0$. Well, that is certainly not true. That is a contradiction; that is an absurdity.

What happened then was, by assuming there was a solution to that equation, I have reasoned my way down and reached a contradiction. That means, I can conclude that no solution exists to that equation. That is an indirect proof. It calls for a Q.E.D. Proof is over. So, we will see these kinds of arguments in the course.

Indirect proof is really very slick. The mathematician G. H. Hardy described it in a very charming way. He was thinking of chess where you make a sacrifice, you give up some chess piece in the hopes of eventually winning. What he did was, he saw proof by contradiction as being rather the same thing. You give up the result you want to prove. You say, all right, maybe it is false. You are going out on a limb. Here is what Hardy said, the weapon (of proof by contradiction) is "…a far finer gambit than any chess gambit: a chess player may offer the sacrifice of a pawn or even a piece, but a mathematician offers *the game*." Proof by contradiction, you give away your conclusion, but it is a fake sacrifice because in the end, you establish it.

That is one bit of logic, the different proof strategies. The other bit of logic I want to mention here is the contrast between a statement and its converse. This will be important throughout the course. Let us just review that. A statement will have the form if A, then B. The converse reverses the role of the hypothesis in the conclusion. So instead of if A then B, if hypothesis then conclusion, you then flip it around, the converse says, if B then A. This is an interchange. It forms these logical cousins if you will, the statement and its converse. Mathematicians, if they have proved a statement, almost always say: What about the converse? Let's take a look at it. It makes a nice logical package.

Let me give an example of a statement in converse from the realm of plane geometry. Here is my theory: If a triangle has three equal sides, it has three equal angles. You have a triangle, it is equilateral, three equal sides. Does it follow that it has three equal angles? Well, if you remember from your geography course, it does. So that is true. That theorem is correct. We would say it this way: An equilateral triangle is an equiangular triangle.

What about the converse? The converse would be, remember you have to flip the hypothesis and conclusion. Now it would say, if a triangle has three equal angles, it has three equal sides. If it is equiangular, then it is equilateral. Yes, that is true also. Here is a case where the theorem is true and its converse is as well. That is great. Mathematicians love that sort of thing. I should point out that it does not necessarily follow that the converse must be true just because the theorem is. This is an important fact to remember. The converse need not be true. It might be, as it was in this case. It does not have to be.

To illustrate that, I will take an example from the world of pets. Suppose I have this theorem: If Fido is a dog, then Fido is a mammal. Well sure, if Fido is a dog, Fido is a mammal; that is true. What is the converse to that? The converse would say if Fido is a mammal, then Fido is a dog. When you think about that, you see that is not true. I might have a pet. My pet Fido might be a mammal, but it might be a wombat or something. It does not have to be a dog.

There is a case where the converse is not true. If mathematicians prove a theorem, they have to start afresh to prove the converse. It does not automatically follow. There is one great example we will see of this in a few lectures, and that is the Pythagorean Theorem. Everybody knows the Pythagorean Theorem. If you have a right triangle, then the square on the hypotonus is the sum of the squares on the other two sides. That is as great a theorem as there is.

What about the converse? If you have a triangle where the square on one side is the sum of the squares on the other two, then it is a right triangle. Is that true? It does not have to be. We have seen the converse does not automatically follow. That is a separate problem. A lot of people who know the Pythagorean Theorem do not know whether the converse is valid or not. Euclid checked this out. It turned out to be the last proposition in the first book of Euclid's *Elements*. We will see him attack the converse. So stay tuned.

I want to conclude this lecture by just giving you a brief introduction to some of the great thinkers whom we will meet here. There is quite an array of spectacularly good mathematicians. Let me just mention a few of them.

One of them is Euclid; I just mentioned Euclid and his *Elements*. I would show you a picture of Euclid, but nobody knows what Euclid looked like. We do not have likenesses of these ancient folks. If you see a picture of Euclid, it is some artist's rendition. Euclid's picture will be somebody with a beard and a toga. That is about it. He kind of looks like pictures you see of Homer or Aeschylus. They are all the same pictures. We really do not know what they look like.

Later, in the course I should say, we will get to mathematicians that we do know what they look like because we have paintings of them. By the end we will actually have photographs, which is sort of neat; but for Euclid, we do not know what he looked like but we sure know what he wrote, and that was the *Elements*. Euclid's *Elements*, pretty much universally regarded as the greatest mathematics text ever. We are going to spend three lectures looking at this and trying to see why this work is so famous, so revered.

I have up here on the screen a Frontispiece from an English translation of the *Elements* from 1570. Very long ago this was an important text. They would make these books with these wonderful decorated opening pages like this one. We will meet Euclid.

His work, the *Elements*, was so famous that you did not even have to mention that you were talking about Euclid's *Elements* if you would just cite a number, a theorem number, people would automatically know what it was. I liken it to this, if I said I am going to read to you Genesis 2:3, I do not have to say, oh, this is from the Bible. Everybody knows that, it is so famous. If I said my First Amendment rights are violated, no one is going to say, First Amendment to what. They know, First Amendment to the US Constitution. For so many centuries in mathematics, if you said, "that theorem follows from I.47," what that meant was the 47th proposition of the first book of Euclid and you did not even have to say, Euclid's *Elements*. I.47 was good enough, this result was that famous.

Who are some other great thinkers? Well, one is Archimedes. What did Archimedes look like? He looked like Euclid. I am pretty sure of that. We are going to take a look at a proof of Archimedes that reads as follows: "The area of any circle is equal to a right-angled triangle in which one of the sides about the right angle is equal to the radius, and the other to the circumference, of the circle." Now, this looks quite convoluted if you read this. It is just a bunch of words; it is hard to see what he is getting at. The Greeks did not have Algebra. If they wanted to express a result that we would write as a formula, they had to write it as a sentence like this; some of which end up sounding a little complicated, some of which actually end up sounding rather poetic.

This is a great theorem. What it is, if you translate it into modern terminology is, it is looking at a circle and telling you what the area is, $A = \pi\, r^2$. That result is due to Archimedes. We will see it proved. This is the one that uses the double contradiction. It is a great argument. He was a great thinker, if ever there was one.

The great mathematicians were so extraordinary that G.H. Hardy said, "Greek mathematics is the real thing." This is not some primitive precursor. This is real mathematics. Hardy went on to write this, "The Greeks first spoke a language which modern mathematicians can understand; as Littlewood said to me once, they are not clever schoolboys or 'scholarship candidates,' but 'Fellows of another college'." The Greeks are your colleague if you are a mathematician. We will spend a good bit of time with the Greeks.

Then we will jump to Medieval Islam, to the world of Muhammad ibn Mūsā al-Khwārizmī, the great Arabic scholar who wrote a treatise on Algebra that is regarded as the greatest mathematics text of the medieval period, and was very influential not only in the Islamic world, but in Europe after the Renaissance.

We will meet Gerolamo Cardano, this very strange character from 16th Century Italy who published the first proof of the solution of the cubic equation, the third degree equation, by subdividing a cube in a most fascinating way. It is a great proof from a very strange person. I think you will enjoy meeting Cardano.

In the next century, the 17th, we will meet Newton and Leibniz; Isaac Newton, Gottfried Wilhelm Leibniz, the co-creators of the calculous. We'll talk about their extraordinary, genius and their unseemly battle with one another over which of them deserved more credit for this great subject.

Moving on, we will run into Leonhard Euler, the great Swiss mathematician, the most prolific mathematician in history, and his successor Carl Friedrich Gauss. He was not so prolific, but every bit as powerful. We will end up with Georg Cantor in the 19th Century who gave us the theory of the infinite, a very profound, radical, exciting idea. Those will be some of our great thinkers.

Let me end this with a couple of quotations from Bertrand Russell. Russell was a mathematician, a logician, a philosopher, a social critic, winner of the Nobel Prize for literature. He was quite an extraordinary character of his own. He characterized mathematics in these words, "… knowing no compromise, no practical limitations, no barrier to the creative activity embodying … the passionate aspiration after the perfect from which all great works springs." This was his characterization of mathematics.

Russell went on to say, "Remote from human passions, remote even from the pitiful facts of nature, the generations have gradually created an ordered cosmos where pure thought can dwell as in its natural home." I love that last line. Russell had a way with words, but I guess that is what it takes to win the Nobel Prize in literature. These sentiments will guide us in the lectures to come as we consider our theorems as works of art, occupying that place where pure thought dwells.

With that sendoff, we will begin our long journey through the history of mathematics. Bring on the great thinkers. Bring on the great theorems.

Mathematics before Euclid
Lecture 2

**Let me quote Richard Trudeau, a math author who said this: "…
when the pall of familiarity lifts, as it occasionally does, and I see the
Theorem of Pythagoras afresh, I am flabbergasted." It would be nice
to try to retain that sense of wonder as we approach this great result.
It is quite amazing.**

Before we launch into Euclid's *Elements*, we take a brief look at the
robust mathematical traditions of three non-Greek civilizations,
Egypt, Mesopotamia, and China. We also meet two Greek
mathematicians who predated Euclid, Thales and Pythagoras.

The Moscow Papyrus, preserved from ancient Egypt (c. 1850 B.C.), contains
a theorem for finding the volume of a particular frustum of a pyramid.
(A frustum is the object formed if we imagine slicing off the peak of a
pyramid with a plane that is parallel to the pyramid's base.) Three important
observations have been made about this papyrus: (1) The original scribe had
to guess at the correct answer for this particular frustum. (2) The idea of
attacking a general problem rather than a specific one seems to have been
foreign to the Egyptians. (3) The papyrus includes no suggestion about why
the given formula works; it is presented merely as a recipe.

The civilization of Mesopotamia, like Egypt, also had a rich mathematical
tradition, often focused on the properties of whole numbers. Much of the
mathematics of Mesopotamia is known to us from clay tablets that have
survived through the centuries. A document called the *Chou Pei Suan Ching*
gives us an example of ancient Chinese mathematics.

The idea of proving a general mathematical result originated with the Greeks,
and according to tradition, the first mathematician to prove theorems was
Thales. Among the theorems attributed to Thales are the following: (1) The
base angles of an isosceles triangle are equal (congruent), (2) the sum of the
angles of any triangle is two right angles (180 degrees), and (3) an angle
inscribed in a semicircle is a right angle.

Of course, Pythagoras is another well-known mathematician from before the time of Euclid. He was born on the island of Samos but later moved to Crotona in southern Italy to establish a kind of think tank or early university. There, Pythagoras and his followers studied music, astronomy, and mathematics, believing that an understanding of mathematics would lead to greater understanding of the world.

The idea of proving a general mathematical result originated with the Greeks.

The Pythagorean theorem states that the square on the hypotenuse of a right triangle is the sum of the squares on the other two sides. We write this as $c^2 = a^2 + b^2$, but the Greeks did not think of the theorem as an equation with exponents; they thought of it as literal squares—squares built on the hypotenuse and the legs of the triangle. Looking at this idea pictorially, we can see that the area of the square built on the hypotenuse is the sum of the areas of the squares built on the legs. In other words, the Greeks thought about this theorem in terms of areas of squares.

It's likely that Pythagoras proved his theorem by starting with two squares of the same size, then dividing the squares up strategically and adding the areas of the resulting shapes. Dividing the squares differently, we get $a^2 + b^2 +$ four triangles $= c^2 +$ four triangles; we then subtract the triangles to get $a^2 + b^2 = c^2$. If we don't like algebra, we can also prove the Pythagorean theorem geometrically. In fact, there are hundreds of proofs of this theorem, many collected in a book called *The Pythagorean Proposition* by Elisha Scott Loomis. Just as artists are compelled to paint different landscapes, so are mathematicians compelled to make their mark by proving this result in unique ways. ∎

Suggested Reading

Heath, *A History of Greek Mathematics*.

Joseph, *The Crest of the Peacock*.

Katz, ed., *The Mathematics of Egypt, Mesopotamia, China, India, and Islam*.

Loomis, *The Pythagorean Proposition.*

Robson, *Mathematics in Ancient Iraq.*

Swetz and Kao, *Was Pythagoras Chinese?*

Questions to Consider

1. For $\triangle ABC$, suppose AC and BC are of the same length and that the (degree) measure of $\angle ACB = 20°$ Suppose further that AD bisects $\angle CAB$ and BE bisects $\angle CBA$, with the two bisectors meeting at F. Using results attributed to Thales, find the degree measure of $\angle AFB$.

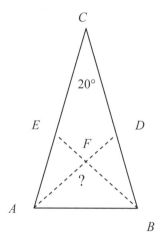

2. Here's a problem from an ancient Chinese text: A bamboo shoot 10 ch'ih tall is broken. The main shoot and its broken portion form a triangle. The top touches the ground 3 ch'ih from the stem. What is the length of the stem that is left standing erect? HINT: As noted in the lecture, the Chinese knew the Pythagorean theorem.

3

Mathematics before Euclid
Lecture 2—Transcript

The first great work we will examine in this course is Euclid's *Elements*, which is a masterpiece indeed. That will begin in our next lecture. In this lecture, I want to talk about some pre-Euclidian mathematics. In particular, I want to give a brief introduction to the mathematics of three non-Greek civilizations that had robust traditions, and then introduce a pair of Greek mathematicians who predated Euclid, but who left very deep footprints.

The first of our non-Greek civilizations is that of ancient Egypt. We are back at the dawn of recorded history. Our knowledge of their mathematics comes from papyri, papyrus roles that have survived all of these years. In particular, there is one called the Moscow Papyrus dated around 1850 BC. This contains a theorem, a result, about pyramids. We know that Egyptians loved their pyramids. Actually it is not a full pyramid, but something called the frustum of a pyramid, and that might require just a little review.

What is a frustum? Here is the idea. If you have a pyramid with a square base coming to a peak, and you take a plane parallel to that base and sliced through it, on top you would have a little pyramid. Remove that and what is left on the bottom is called the frustum. It is like a pedestal. Here is a picture of one. In fact, this is the very one that is going to be referred to in the Moscow Papyrus.

If you look at the dimensions, the square base around the bottom is 4 × 4 × 4 × 4. It tapers up to a square top that is 2 × 2 × 2 × 2. The gap between the top and bottom is six units. This is the frustum that we are going to be looking at.

What do we do with it? The question is find its volume. Find its volume? That is not trivial. This is what the Egyptian scribe writes about it in the Papyrus. Let me read to you the solution, how you find the volume of this thing. Here is what the Papyrus says. You are to square this 4, result 16. You are to double 4, result 8. You are to square 2, result 4. You are to add the 16, the 8, and the 4, result 28. You are to take one-third of 6, result 2. You are to take 28 twice, result 56. See, it is 56. You will find it right.

A modern person looking at this is completely befuddled. What is this about? Where are these numbers coming from? What is going on here? It is very opaque. It requires three observations. First of all, guess what the right answer is. The volume of that frustum is 56. They got it. The Egyptians got it. I would hate to think what would happen if you went out on the street today and asked somebody to find the volume of that frustum. How many would come up with 56, as did the Egyptian scribe who wrote this Papyrus almost 4,000 years ago. That is pretty impressive. They must have been onto something.

Second observation is that the Papyrus addressed a particular frustum; just that one with those dimensions 4 × 4 × 4 × 4, 2 × 2 × 2 × 2, height 6. That was fine, but what if you have a frustum that is 5 × 5 × 5 × 5 and 3 × 3 × 3 × 3, and height 12. What do you do? What you have to do is look at the recipe that they gave you for their frustum and try to make the analogy between their numbers and the ones you have, the dimensions of your solid, and figure out how to make this conversion so that you get the right answer following their recipe. It is not spelled out in a general fashion. It is just that one particular body that they are looking at.

We would expect that this should be done in general. We would imagine a frustum that was $a \times a \times a \times a$, the square on top, $b \times b \times b \times b$, the height between them h, and we would come up with a general formula, which is this. This is the actual answer to this thing. The volume is $\frac{1}{3} h (a^2 + ab + b^2)$. You could use this for any frustum you encounter. You are not restricted to that particular one.

This idea of attacking a general problem rather than a specific one was foreign to the Egyptians. It is now something of course we have come to expect. I should note if you use that general formula and put in the dimensions of their frustum, $a = 4$, $b = 2$, $h = 6$, you get 56. Naturally, you get the right answer.

The third observation is this. This is maybe the most critical. There is no suggestion in this Papyrus as to why this works. It is just presented as a recipe, you follow it, and you will find it right you are told, but why. Well, that is not part of the Egyptian tradition.

The second non-Greek civilization I want to mention just briefly is that of Mesopotamia. The Babylonians and their neighbors had a rich mathematical tradition, often focused on numbers, the properties of whole numbers. We know about this through clay tablets that have survived over the centuries. I guess unlike the Egyptians who would write on Papyrus, which if you think about it will decay and dissolve over time, the folks in Mesopotamia would take a stylus and engrave things on a clay tablet which would then either be dried or even baked. It comes out like a brick. These things last a long time. There are museums and libraries all over the world that have clay tablets from Mesopotamia that they are translating and studying to learn about the mathematics of the time. I am not going to say any more about this, because this could be a whole course just on Mesopotamian mathematics, but I wanted to just indicate it was out there and flourishing.

So too was mathematics in China. Again, a whole course could be devoted to it. I am just going to show you one picture, one diagram that we have of Chinese mathematics. This comes from a document called the *Chou Pei Suan ching*. If you look at it, it is a pretty design. You might just enjoy the picture. It has these vertical lines, these horizontal lines forming a grid. In the middle is this sort of cockeyed square sitting there. Why am I showing this design to you? I'm showing it to you because this is doing something mathematical. There is something going on here that is quite sophisticated and quite impressive from the Chinese mathematical tradition. I will ask you to hang on until the end of the lecture to see what it is.

The idea of proving a general mathematical result originated not with any of these civilizations, but with the Greeks. There are two key words there in what I just said: general mathematical result. The Greeks would try to attack the general problem. They would not look at that one frustum. They would look at frusta generally and try to learn something. It was not that individual right triangle of interest, but all right triangles. This is a very important extension to do general results rather than specific ones. Even more important is the word "proving." They wanted to prove the general result, that is give a logical argument starting from easy premises and working your way down until you have established this result for once and for all. The idea of proving things we attribute to the Greeks, and that of course has colored all of mathematics ever since.

The person who is supposedly the first mathematician from ancient Greece, the so-called father of demonstrative geometry is Thales. The legend has it that he was the first person to prove things. Thales lived not in Greece proper, what we think of Greece today, but on the eastern shore of the Aegean which is currently western Turkey. He was renowned for his genius not only in mathematics but also in philosophy and in astronomy. He was one of the wise men of his time. Our interest in Thales is that he is supposedly, according to tradition, the first person to prove theorems. We do not have any documents from this far back, so this is all tradition, but it probably has some soundness.

What theorems, well let me mention three theorems that are attributed to him? One is this very important fact, that the base angles of an isosceles triangle are equal. We would say are congruent, but I will use the old language. The base angles of an isosceles triangle are equal. There is a picture of a triangle ABC. Let's suppose we say it is isosceles, which means that two of the sides are the same, $AC = BC$. Tradition holds that Thales proved from that fact that $\angle A = \angle B$. The base angles of an isosceles triangle are equal, famous result, Thales gets credit for it.

The second result, the sum of the angles of any triangle is two right angles. This is another one of the theorems that Thales supposedly addressed. We take a triangle, any triangle ABC, and what the theorem is saying is that if you take $\angle A + \angle B + \angle C$ = two right angles. You might expect that to say 180 degrees, but at this time the Greeks did not measure angles in degrees. Rather, you just measured them in terms of how many right angles there were. For a triangle, the answer is two, from Thales.

Finally, the third result I am going to mention is called Thales' Theorem, a bit of a tongue twister there. This is less well known and I am going to actually show you a proof of this one. It is very nice. It says an angle inscribed in a semicircle is right. What does that even mean? Here is what's up. Suppose I have a semicircle. My center is O, my radius is OA, I draw the semicircle, any semicircle, this is generic, then you pick a point C anywhere on the semicircle. The C is also entirely general. Draw a line AC, draw a line BC, and you have thereby constructed an angle inscribed in a semicircle. That is $\angle ACB$ is inscribed within the semicircle. What Thales theorem contends and will prove is that is a right angle, $\angle ACB$ is a right angle. Why is that? Let's see a proof.

There is my diagram. This proof, again, we are not sure this is what Thales did, but everybody thinks it is. It is such a beautiful, natural argument. It is actually what Euclid uses in Book III of the *Elements* when he proves this theorem. This is probably what Thales did.

Let me look at triangle *ABC* and label the angles alpha and beta, the angles at *A* and *B*. We are going to keep track of those. The critical step is this, draw line *OC*. That is, you draw a line from the center of the semicircle out to your point on the semicircle, *OC*. That creates two triangles. You can see them there, left and right, and we are going to examine each of those two triangles individually.

First of all, look at triangle *AOC*, the one to the left. Notice that *OA* and *OC* are equal because they are radii of the semicircle. All radii are equal. Triangle *OAC* there is isosceles. It has two equal sides and hence two equal angles by the result of Thales, meaning that $\angle ACO$ is also alpha. The base angles of the isosceles triangle are the same. I can put an α up there at $\angle ACO$. Triangle *BOC* to the right is also isosceles for the same reason, *OB* is a radius, *OC* is a radius. You have a triangle with two equal sides hence the base angles are equal. If there is a beta at one base angle, there is a beta up there at $\angle OCB$. You have kind of distributed the angles around and now let us just look at the big picture.

In triangle *ABC*, the big triangle we formed, I know that $\angle A + \angle B + \angle C$ at the top must add up to two right angles. That is what Thales supposedly had proved, but $\angle A$ is α, $\angle B$ is β, and $\angle C$ up there = $\alpha + \beta$. That would add up to two right angles. Look at the left side, $\alpha + \beta + (\alpha + \beta) = 2\alpha + 2\beta$ is two right angles. Split everything in half, $\alpha + \beta$ is one right angle, but $\alpha + \beta = \angle ACB$, so $\angle ACB = \alpha + \beta$ which is a right angle. That is what we had to prove. The angle inscribed in the semicircle is right, QED, the proof is over. This is Thales' theorem, a nice result from long, long ago in Greek mathematics.

The other mathematician from before Euclid that I want to mention is Pythagoras. Everyone has heard of Pythagoras. He was born on Samos, which is an island just off the coast of modern day Turkey in the eastern Aegean. That is not very far from where Thales lived, and it is possible that a young Pythagoras might have even met an old Thales.

In any event, Pythagoras rather quickly pulls up stakes and moves to southern Italy to a place called Crotona and takes with him his followers, the so-called Pythagorean Brotherhood where they establish what you could imagine to be a kind of think tank, or even an early version of a university. Pythagoras and the followers think together, work together, contemplate nature together, and make great progress. They studied music, they studied astronomy and they studied mathematics. The Pythagorean philosophy was this: If you could understand mathematics the way mathematics works, you can understand the world beyond. You can apply mathematics to all sorts of different things out there in the real world. There's an idea that has caught on; to this day mathematics is as applied as it gets. You can apply mathematics to all sorts of things. If you want to trace back the philosophical origins of that approach, you can trace it back to the Pythagoreans.

It was probably never better stated than by Galileo many centuries after Pythagoras, but also working in Italy. Galileo wrote this: "The universe is … a grand book that cannot be understood unless one first learns to comprehend the language and read the letters in which it is composed. It is written in the language of mathematics." That idea is still with us. That idea goes back to the Pythagoreans.

It is not for that that I want to mention Pythagoras, but rather for the great theorem that bears his name, the Pythagorean Theorem. Remember what it is, you have a right triangle, here is your right triangle, with legs of length a and b, hypotenuse c and I have my two angles labeled there, the angle opposite side a I will call alpha. The angle opposite side b I will call beta. This is a right triangle. That is what the Pythagorean Theorem is about. What the theorem says is, if you take that right triangle, the square on the hypotenuse is the sum of the squares on the other two sides. Everybody knows that.

What not everybody does know however is that the Greeks thought of this not as an equation with exponents—we write the Pythagorean Theorem as $c^2 = a^2 + b^2$—they thought of it as literal squares; squares built upon the hypotenuse, squares built upon the legs. If you look at the picture, I have done this. I have built this green square on the hypotenuse, a blue square and a grey square on the legs. What the theorem says is the area of the green square on the hypotenuse is the sum of the areas of the blue and the grey squares on the legs.

It is quite phenomenal. It is about areas of squares, not our familiar equation $c^2 = a^2 + b^2$. Why not? Well, the Greeks did not have Algebra. They could not have written that. They would not have known what that meant. They did not have exponents. That is a very concise modern way of writing the Pythagorean Theorem. The Greeks thought about this in terms of areas of squares.

We are so familiar with this result. Pythagorean Theorem, you learned it in school, you use it. If you know Trigonometry, it is the result behind Trig. It becomes second nature. It is easy to forget how amazing this is. Why is it that the square on the hypotenuse of a right triangle is the sum of the squares on the legs? It is not at all self evident that this is going to work that way. It is really a result where familiarity can breed contempt, or at least indifference. We should look at this and realize how spectacular a result it is. It is as great a theorem as mathematics can boast.

Along those lines, let me quote Richard Trudeau, a math author who said this, "… when the pall of familiarity lifts, as it occasionally does, and I see the Theorem of Pythagoras afresh, I am flabbergasted." It would be nice to try to retain that sense of wonder as we approach this great result. It is quite amazing.

What I would like to do is show you a proof of the Pythagorean Theorem. If it is this important of a result, we had better prove it in course about great theorems. This proof is one that is often attributed to Pythagoras. This conceivably is how he would have done it. Again, we are not sure. We do not have documents from that ancient time. We are kind of guessing, but there is a certain naturalness to this argument that would suggest that this might be the way it went.

What do we do? Remember we have our right triangle of sides a, b, and c. What I am going to do first is build two big squares of the same size; side $a +$

$b, a + b, a + b, a + b$ on the left-hand square, and the same dimensions on the right. Each of these squares is of side $a + b$. Obviously, they have the same area to start with. What I am going to do now is divvy these up, break these up in different ways and see what happens.

Look at the left-hand square. Starting in the upper left-hand corner, I am going to mark off along the top a length b. If the whole length across the top was $a + b$ and I put a b there on the left-hand part, then the right-hand part up across the top must be a. There is a segment of length b and then a segment of length a across the top.

Coming down the right side, I am going to first mark off a segment of length a, and then what is left would be of length b. Going across the bottom, I am going to pick up an a and then a b. Rising up the left side, a b and then an a. This is a very strategic way of breaking this up, of locating these points. Just to be clear, if I were to start in the upper left and go around clockwise, I am going to encounter ba, ab, ab, ba. Connect these points with a vertical and horizontal line, and split the rectangle in upper left and lower right with diagonals. I put in a diagonal in the upper left, and one in the lower right. I have broken this left-hand square into that picture.

What is the area of that left-hand square, the big one? What I am going to do is just add up the little pieces into which I have split it. Notice we have in the upper right a little square $a \times a$. I put in its area that is a^2. That is part of the area of the big left-hand square. In the lower left we have a square also, $b \times b$. That would give me a b^2 down there. What's left then are four copies of our original triangle. Can you see that? Those four triangles that have been formed by the diagonals each have a side a, a side b, and a right angle. They are all copies of what we started with. I am going to say that the area of the left hand square = $a^2 + b^2 + 4$ (triangles). Just to see them, how about we will color them in green. There they are. That is one way of thinking of the area of the left-hand square with that sort of subdivision. We will store that for a minute. We will be back to it.

Now I want to look at the right-hand square, which remember was the same size as the left-hand square. I have shown my triangle again, side a, side b, hypotenuse c. Now I am going to mark off points around the right-hand square, the big square in this fashion. Across the top I mark off a chunk of size b, and what remains then will be of size a. Down the right, the same thing, b first, then a; across the bottom, b first, then a; up the left side, b first, then a.

If you do the trip around this square starting in the upper left, you hit ba, ba, ba, ba. It is a different subdivision. What I will do is connect those points I have thus formed, so I get four lines inside. There they are. That creates this quadrilateral inside the big square. Notice now what I have are triangles in each corner. Those triangles again are exact copies of the one I started with, side a, side b, a right angle. These are identical. The diagonal lines, the oblique lines, those four oblique lines within the big square, each is of length c. I have labeled those.

The temptation now is to say, oh that is a square inside there because it is $c \times c \times c \times c$. That does not make it a square. A figure can have four equal sides but not be square. Think of a diamond, which has four equal sides but is stretched out. I have more to do if I claim that is really a square in there. I have to make sure it has four right angles.

How does that work? Let me call the angle up at the top, which I have just entered there, call that gamma. What my goal is is to convince you that is a right angle. How do you do that? Well, look at the upper left-hand triangle. It is a copy of my original, and so the angle just to the left of gamma must be an alpha. Look at the original triangle. The angle opposite side a is alpha. There it is. I will label that alpha.

Look at the triangle on the upper right. It too is a copy of my original, and so the angle just to the right of gamma is going to be beta. It is the angle opposite side b. Along that horizontal line up at the top, I have an angle of alpha, then an angle of gamma, then an angle of beta. That is going to certainly figure into where we are going. Look now at the right triangle that we started with. It has three angles, an alpha, a beta, and right angle. You know that the sum of the three angles of a triangle is two right angles, so the $\alpha + \beta$ must add up to a single right angle. That is true of any right triangle. The sum of the two acute angles is a right angle.

Now we can put all of this together and reach my conclusion. If I look at that horizontal line at the top of the right hand square with the alpha, the gamma, and the beta as the three angles around that line, I know that the sum of the angles around the line is two right angles. Two right angles is $\alpha + \beta + \gamma$. You just add those three up. We just said $\alpha + \beta$ is a right angle, so I replace $\alpha + \beta$ with a right angle, and thus conclude that two right angles is one right angle plus gamma. Well, subtract a right angle from each side, gamma must be a right angle. That is what I wanted to show. In fact, so are the other three angles of that interior quadrilateral; they are all right angles. My quadrilateral in there has four equal sides, four right angles; bingo, it is a square. That is the critical step to finish the proof.

Here is how we will do that. I will put both of my big squares up there. There is the left-hand square, the right-hand square subdivided differently. Remember we showed the area of the left-hand square was the $a^2 + b^2 +$ the four triangles. There they are colored in. The area of the right-hand square, now what is that going to be, the one on the right? Well, fist of all, it has this square in the middle, kind of cockeyed, but it is $c \times c$, so that is a c^2 is the area of that square. The only other part there is the four triangles around the edges, and they are all copies of my original triangle. I get four triangles added to the c^2.

You say this, look, the left-hand square and the right-hand square were equal in area to start with, the left hand square is $a^2 + b^2 + 4$(triangles). The right-hand square is $a^2 + b^2 + 4$(triangles), so $a^2 + b^2 +$ four triangles $= c^2 + 4$ (triangles). If you subtract the four triangles from each, you get that $a^2 + b^2 = c^2$. My picture then would lead to this conclusion $a^2 + b^2 = c^2$, the Pythagorean Theorem would be proved. You can see it in this picture.

If you do not like the algebra, I am going to show you a way to prove this geometrically by just removing the green triangles. They are all the same area, so stare at the picture, left-hand square, right-hand square and at the count of three, the green triangles go, 1, 2, 3, bingo. There it is. The square on the hypotenuse is the sum of the squares on the legs. You can see it jumping out at you. That is the Pythagorean Theorem. We express it as $a^2 + b^2 = c^2$. I like the picture to prove it for us. There is the proof; maybe the proof due to the Pythagoreans.

You know what I hope you notice is that the picture we saw along the way looks familiar. Do you remember the Chinese diagram we saw from long ago and from a very different part of the world, there it is; the cockeyed square within the square. What this diagram is doing is proving the Pythagorean Theorem, the Chinese style. It is looking at a right triangle of sides 3, 4, and 5 and demonstrating that the square on the hypotenuse is the sum of the squares on the legs. That is curious. You see the same picture popping up in two different cultures. The poets in China and Greece were both writing about love, the mathematicians in China and Greece were both discovering this picture. It suggests the universal nature of mathematics.

I should mention a quirky book called *The Pythagorean Proposition* by Elisha Scott Loomis. He published this about a century ago. In this he has collected hundreds of proofs of the Pythagorean Theorem, hundreds of them; not just this one, but there are lots and lots of alternatives. It is kind of crazy to see so many proofs of one result, and you might ask why do we need more than one proof of a result; you do not. One proof establishes it.

You wouldn't ask why do you need just one landscape, why do you need just one portrait. Different artists want to do their own landscapes, their own portraits. Different mathematicians have tried to do different proofs, unique proofs of the Pythagorean Theorem, and did they ever. There are more proofs of the Pythagorean Theorem than of any other result in mathematics. If you are interested, you can read Loomis' book and find proofs of the Pythagorean Theorem to your heart's content.

With this, the stage is set. We have seen the pre-Euclidian mathematics, now it is time to get on to the great one, Euclid and his *Elements*.

The Greatest Mathematics Book of All
Lecture 3

---

*After you've done the first four propositions, proposition I.5 is a little bit more challenging, and some people, namely, the mathematical asses, could not cross this bridge, could not get over the **pons asinorum** and, thus, enter the remainder of Euclid's Elements.*

W e know little about the life of Euclid, except for the fact that he was the leading mathematician at the great Library of Alexandria in Egypt. Somewhere around 300 B.C., he wrote the *Elements*, a vast compendium of mathematics broken into 13 books and containing 465 propositions, or theorems. The *Elements* was highly regarded throughout antiquity and even up to the 19th and 20th centuries, when it was studied by such thinkers as Abraham Lincoln and Albert Einstein.

Euclid begins with some definitions, some postulates, and some "common notions," as he calls them. The postulates and common notions are self-evident truths, not requiring proof. Then, using the definitions, postulates, and common notions, along with reason, Euclid deduces a consequence, called the first proposition. From there, he deduces a second proposition and so on, building on his foundation of axioms and definitions.

Some of the terms Euclid defines are quite familiar to us, such as an isosceles triangle and a circle. One unusual definition is number 10, in which Euclid defines a right angle using perpendicular lines rather than degrees. Another interesting definition is number 23, in which he defines parallel lines as being in the same plane but never meeting. Euclid does not say that such lines must be everywhere equidistant or have the same slope.

Euclid next defines five postulates: (1) It is possible to draw a straight line from any point to any point. (2) It is possible to produce a finite straight line continuously in a straight line. (3) It is possible to describe a circle with any center in any radius. (4) All right angles are equal. (5) When a straight line falling on two straight lines makes the interior angles on the same side less than two right angles, the two straight lines, if produced indefinitely, meet on that side on which the angles are less than the two right angles. A picture helps us visualize this last postulate. Following the postulates, Euclid states his five common notions, which deal with equalities and inequalities.

Euclid's first proof, called proposition I.1, is that an equilateral triangle can be constructed on a finite straight line. This proposition can be demonstrated graphically, using the postulates and common notions. Proposition I.4 is called the side-angle-side congruence scheme; it's a way to show that triangles are congruent. Proposition I.5 proves that the base angles in isosceles triangles are equal. Proposition I.8 is the side-side-side congruence scheme: If the three sides of one triangle equal, respectively, the three sides of another, then the triangles are congruent. Proposition I.11 shows how to draw a straight line at right angles to a given straight line from a given point on it. Again, Euclid shows how to construct the perpendicular line using only the definitions, postulates, and earlier propositions. Finally, proposition I.20 is the triangle inequality; it states: In any triangle, two sides taken together in any manner are greater than the remaining side. Epicurean philosophers later questioned the need to prove this proposition, but a Greek commentator named Proclus defended Euclid: "Granting that the theorem is evident to sense-perception, it is still not clear for scientific thought. Many things have this character; for example, that fire warms. This is clear to perception, but it is the task of science to find out how it warms." ∎

Suggested Reading

Euclid, *Euclid's Elements*.

Heath, *A History of Greek Mathematics*.

Loomis, *The Pythagorean Proposition*.

Proclus, *A Commentary on the First Book of Euclid's Elements*.

Questions to Consider

1. Get a copy of Euclid's *Elements*, Book I, proposition 5 (or find it online at http://aleph0.clarku.edu/~djoyce/java/elements/bookI/prop15. html). This is his proof that the base angles of an isosceles triangle

are congruent. It immediately follows I.4, which is the side-angle-side congruence scheme (SAS). Read the proof, master it, and thereby cross over the *pons asinorum*!

2. Which side do you support in the Euclid/Epicurean controversy (proposition I.20)? Was Euclid right (as Proclus contended) in *proving* the triangle inequality, even though it is "evident to an ass," or could Euclid have glossed over such trivialities? (Related question: Is the "trivial" really trivial?)

The Greatest Mathematics Book of All
Lecture 3—Transcript

This is the first of three lectures I will be giving on Euclid's *Elements*. The greatest mathematics text of classical times; I would argue the greatest mathematics text of all time.

What I want to do is give you an overview of the initial parts of Book I of Euclid's *Elements*, how he gets started, take a look at some of the early results. But before that, I want to give a sense as to why this work has meant so much to so many over so many centuries. I have got to sell the idea of devoting three lectures to this great classic text.

Let me begin with the observation that of Euclid's life we know virtually nothing. He ended up in Alexandria, Egypt heading the School of Mathematics at the great Library of Alexandria. You recall that Alexander the Great had conquered the Mediterranean world, and at the mouth of the Nile had set up this great library to be the center of knowledge and scholarship. Reports are that at its height it had hundreds of thousands of volumes, although volumes in those days weren't books but papyrus rolls.

Euclid was brought down to be the head of the School of Mathematics and he sure did a good job because the mathematics school at Alexandria was preeminent in the world for centuries after Euclid's time.

While there, somewhere around 300 BC, he wrote the *Elements*, his great masterpiece. This is a vast compendium; it's broken into 13 books, as he calls them, or chapters we would say, that contain altogether 465 propositions, or theorems, that he proves. It's a very big work. I liken it to the *Iliad* of Homer. I think the *Elements* of Euclid and the *Iliad* of Homer are cousins, if you will, from two different disciplines. They're both old, they're both big, no one reading the *Iliad* would confuse it with modern literature. No one reading the *Elements* would confuse it with modern mathematics. Euclid's *Elements* has no functions in it, it had no trigonometry, it has no calculus. It's old math.

But anyone reading the *Iliad* would agree that this is great literature. It's important, it's powerful, it's beautiful, and I would say anyone reading

Euclid's *Elements* would recognize this as great mathematics, powerful, beautiful, important. So these, I think, we can keep as a tandem, the *Elements* and the *Iliad*.

In classical times the *Elements* was highly regarded. People like Archimedes, a few generations after Euclid, already were referring to it. Heron referred to it, Cicero referred to it. When classical civilization fell and Greek scholarship was transported to the Islamic world, the *Elements* went with it, and it was highly revered by the Islamic scholars who translated it and studied it.

When the renaissance comes back to Europe, the *Elements* is brought back to Italy, Sicily, places like that, where it's translated into Latin and has a sort of second birth in Europe. In the great universities of the day it was studied. Newton read Euclid's *Elements* and thought very highly of it. In the Enlightenment it was an important document, Jefferson, Ben Franklin, these folks were quite familiar with it; and even in the more modern times it had still had its impact.

I like the story from the 19th Century of a young, Illinois lawyer who was trying to sharpen his reasoning skills. He thought that as a lawyer he would have to prove points of law to judges and juries, and he wanted to get better at this. His name was Abraham Lincoln and he tells us this story. He said, "I left my situation in Springfield, went home to my father's house and stayed there till I could give any proposition in the first six books of Euclid at sight. I then found out what 'demonstrate' means, and went back to my law studies."

Once Lincoln had read the master, Euclid, he knew how to prove something in a clear, rigorous fashion. It is often said that Lincoln's prose reflects his interest in the Bible and Shakespeare, and that is certainly true. You can hear the cadences of those great works in his masterful prose.

But I would argue you can also see the echoes of Euclid in Lincoln's writing as Lincoln will lay out a few hypotheses and then deduce consequences of them in a way that just seems absolutely clear and perfect. So, Lincoln was a fan. Later in the 19th Century we hear another testimonial, this one from Bertrand Russell, a great philosopher. He said, "At the age of 11 I began Euclid with my brother as tutor. This was one of the great events of my life, as dazzling as first love." How often do math books get referred to like that?

Even into the 20th Century we have this passage, "If Euclid failed to kindle your youthful enthusiasm, then you were not born to be a scientific thinker." Who wrote that? Albert Einstein.

So, for centuries and centuries this has been revered as a great work, and it's my job now in this lecture and the next two to show you exactly why these folks were so positive about it.

Here's what Euclid does. He begins with some definitions, he's going to tell us what his terms mean, some postulates, and some common notions, as he calls them. The postulates and common notions are self-evident truths, the sorts of things he doesn't have to prove, he's just going to state them. Then, using the definitions, the postulates, the common notions and reason, Euclid will deduce a consequence, and that will be the first proposition. Then, with the definitions, the postulates, the common notions and the first proposition, he'll deduce the second proposition, and the third and the fourth, and on he goes until we get to some very sophisticated mathematics, all of it built upon this foundation of his axioms and definitions.

This is an axiomatic development. This is how mathematics is still done, developed from a core of axioms, and you'll find no finer example of it than Euclid's *Elements*.

So, I want to begin, then, with the definitions, so what are these? Well some of the terms he defines are quite familiar to us: an isosceles triangle is one with two equal sides, a circle is a curve, all of whose points are a fixed distance from a given point called the center. No surprises there; but I want to show you two definitions that you might find a little surprising.

One of them is Definition 10, the tenth of his 23 definitions. He's trying to define a right angle in perpendicular lines. If somebody asked me how would I define a right angle, I would say, well, it's 90 degrees of course, but that's not what Euclid could do because he doesn't have degrees. Nowhere in the *Elements*, nowhere in the greatest geometry book ever written, are angles measured by degrees. So he's got to do this some other way, and here's his definition of perpendicular lines in right angles. He says, "When a straight line set up on a straight line makes the adjacent angles equal to one another, each of the equal angles is right and the straight line standing on the other is called a perpendicular to that on which it stands." A picture would help, here's his diagram. He has the straight line CD standing on the straight line AB. We have the two angles adjacent, alpha and beta, and the definition says that if alpha equals beta, then we're going to say angle ADC and angle BDC are right angles, and CD is perpendicular to AB. That's his definition of perpendicularity.

Why that's important is, later whenever he wants to prove something is perpendicular, he must come back to this definition, this is what he means by perpendicular, this is what he's got to do. I'll show you an example where he does just that.

The other definition I wanted to put forth was his last definition, Definition 23, of parallel straight lines. Euclid says that "parallel straight lines are straight lines which, being in the same plane and being produced indefinitely in both directions, do not meet one another in either direction." They are lines that never meet. Now, are they everywhere equal distant? Well, maybe, but that's not his definition. Are they going at the same slope? Well, maybe, but that's not his definition. His definition is they never meet. So if he's going to prove something about parallel lines, he's going to try to show two lines are parallel, he's got to show they never meet.

The definitions being out of the way, it's now time for him to start giving his postulates. There are five postulates. Here they come. The first one is, it is possible to draw a straight line from any point to any point. So you give me two points, you can draw a straight line between them, the Postulate 1.

Postulate 2, it is possible to produce a finite straight line continuously in a straight line. So if you have a little line segment here and you want to extend the straight line in either direction, go ahead. You can do it, cite Postulate 2.

If you remember back to your geometry training, this is exactly what you do with a straightedge. You've got two points, you line up the straightedge and connect them to form a line segment as in Postulate 1, or if you already have a segment, you can kind of line up the straightedge and extend it further. So, these are called the straightedge postulates.

Postulate 3 says that it is possible to describe a circle with any center in any radius, you give me a center, you give me a radius, and I'll make a circle. So give me center, there's a radius, I can make a circle. What's that? That's what you do with a compass. So this is the compass postulate.

Taken together, Postulates 1, 2 and 3 allow you to use compass and straightedge as your tools of construction, and these are called, not surprisingly, the Euclidian tools.

Postulate 4 says all right angles are equal to one another. So what Euclid is saying here is, if you have a line standing on a line so that the adjacent angles are equal and hence, you have formed a right angle, and somewhere over on another part of the plane you have a line standing on a line so that the adjacent angles are equal so that there's a right angle, then he postulates that these two right angles are equal to one another. He doesn't have to prove it, this is just a postulate. He will accept this as a self-evident truth. What this does is, it gives him a uniform standard for the plane, a right angle over here equals a right angle over here, and that will be very important in his development.

The fifth postulate is quite different from the first four, it is much wordier. It says this: When a straight line falling on two straight lines makes the interior angles on the same side less than two right angles, the two straight lines, if produced indefinitely, meet on that side on which the angles are less than the two right angles. Obviously this is way more complex just to state it, it's kind of complicated compared to the nice succinct statement, all right angles are equal. This is a different ball game. It certainly requires a picture.

So let me show you the picture for Postulate 5. We've got a straight line AB, we have another straight line CD, and we have a line falling upon them, EF. I will look at the interior angles on the same side, so I am going to look at alpha and beta there, between the two lines, that's what interior means, and to the right of EF. What this postulate says is, if alpha plus beta, those two angles, sum to less than two right angles, then AB and CD meet to the right on that side. This is sometimes called the parallel postulate. It's called that because Euclid needs this whenever he's going to develop his theory of parallels later in Book I. But that's actually a misnomer, it's not a parallel postulate, we said parallel lines are those that never meet; here he's giving a condition that guarantees that the lines do meet. So, if there was any justice in the world, this should be called the non-parallel postulate.

Well, those are his postulates, the five of them, and then he gives five of what he calls common notions. These are still things he's going to accept without proof, these are pretty general statements such as: Common Notion 1: Things which are equal to the same thing are equal to one another. Common Notion 2: If equals be added to equals, the wholes are equal. Common Notion 3: If equals be subtracted from equals, the remainders are equal. Common Notion 4 says things which coincide with one another are equal to one another. This one is a little peculiar in its content. He seems to be saying this, that if you have two objects which can be made to coincide perfectly so that they are kind of carbon copies of each other, then they are equal in all respects. Their angles are equal, their sides are equal, their perimeters are equal, their areas are equal, they are identical. And finally, Common Notion 5, the last, the whole is greater than the part, he says. This will be his way to get at inequalities, and there are lots of those in Euclid's *Elements*. We are going to see some; inequalities always come out of Common Notion 5.

So, he has got his definitions, he has got his postulates, he has got his common notions. The foundations are ready to go and now he's going to start proving theorems. I sometimes think at this point, if it were I who was doing this, I would freeze. What do you do? Where do you start? What's your first theorem going to be?

Well, fortunately Euclid didn't freeze. Here it is, Proposition I.1 and this is how we usually write it with a Roman Numeral I meaning the first book, and then Hindu-Arabic one meaning the first proposition. So this is the first proposition of the first book, the first thing Euclid proves on a finite straight line he shows you how to construct an equilateral triangle.

Let me take a look at this argument. We'll start with any segment AB. So there it is. His job is to put an equilateral triangle up there. Euclid says, all right, with center A and radius AB, construct a circle. Can he do this? Sure, remember Postulate 3 says if you give me any center in any radius I'll make a circle. So he can cite Postulate 3 to justify that circular arc, so, he started. Now he says, this time let center B be where your circle is centered and AB be the radius, and construct a circle. Again, that's Postulate 3 being invoked, and you get the second circle. Euclid lets C be the point where those two arcs intersect up there. He then draws lines AC and BC, so we draw these two lines. Can you do that? Yes, that's Postulate 1. You give me two points, I'll draw a line between them, which is exactly what he's done here.

In this fashion he has constructed this figure. Now he has to show that it's an equilateral triangle. But that's pretty easy because Euclid now says, look, AC and AB are equal because they're radii of that first circle he drew, that BC and AB are equal because they're radii of the second circle he drew, but if AC equals AB and BC equals AB, then it's also true that AC equals BC by Common Notion 1, things equal to the same thing are equal to each other. So AB equals BC equals AC, all three sides of this triangle are the same, the triangle is equilateral and that's his proof. So, he's got a theorem proved.

I said there's 465 theorems so there's 464 more to do, but we're not going to do them all. So, I'm going to have to start skipping through some of Book I. So, let me hit some high points in the remainder of the book.

Proposition I:4, the fourth proposition is what we call the side-angle-side congruence scheme. It's a way to show triangles are congruent, side-angle-side. You might remember this if you have triangle ABC and triangle DEF, suppose they both have a side A on the bottom, they both have a side of length B to the left, and the angle at A is alpha, the angle at D is alpha. So you have two sides in the included angle of one triangle, equal two sides of the included angle of the other, then we say by side-angle-side the triangles are congruent, meaning they are equal in all respects, they could be made to coincide, their areas are equal, their angles are equal, they are carbon copies. So now that congruent scheme is out there.

Proposition I:5 he proves that in isosceles triangles the angles at the base are equal to one another. This is the result, remember, that Thales supposedly had proved. I'm not going to go through Euclid's proof, but I want to show you his diagram. This is the diagram that accompanies his proof. There you see the isosceles triangles, ABC there. AB and AC are the equal sides, and in the course of the proof he has to do the constructions that you see on the screen. The diagram thus formed has come to be called the *pons asinorum* which translates to the bridge of asses. There is kind of an amusing reason where that name came from. First of all, it sort of looks like a bridge, if you use your imagination you can imagine crossing over on BC, the bridge has the struts below holding it up, so it has a kind of resemblance to a bridge, but more fundamentally it is the bridge to the remainder of Euclid. After you've done the first four propositions which aren't too hard, Proposition I:5 is a little bit more challenging and some people, namely the mathematical asses, could not cross this bridge, could not get over the *pons asinorum*, and thus enter the remainder of Euclid's *Elements*, so this was a barrier for some folks.

Let me jump ahead a few more to Proposition I:8. This is the side-side-side congruence scheme (SSS). If you have two triangles, ABC, DEF, they both have a side A, they both have a side B, they both have a side C, so we have three sides of one, respectively equal to three sides of the other, then these triangles are congruent, identical in all respects. So now he's got two congruent schemes to work with. He's going to use this one in Proposition I:11. I'm going to actually give you this proof; I think this typifies what Euclid is up to.

Proposition I:11, he says, he wants to show you how to draw a straight line at right angles to a given straight line from a given point on it. So, my picture would be this. I have a straight line, there's a point D on it. I want to construct a perpendicular upward at that point. I'm willing to believe there is a perpendicular upward from that point, but I have to create it, I have to construct it. Well, what do you do? Do you get out your protractor? No, you're not allowed to use protractors. That's not one of the allowable tools. Nor can you use your cool software program on the computer. You've got to use compass and straightedge. Euclid has to create this perpendicular with compass and straightedge and prove it is a perpendicular by the definitions that he has given. So, if you think of it that way, this is a little bit more of a challenge than you might think.

Let's see what he does. Euclid says, all right, there's my line and my point D, I want to put up a perpendicular there. He begins by taking a point A on the line somewhere other than D, so I put it to the left of D, so you take a point A. Now what he wants to do is construct a segment going the other way, from D to B, equal in length to DA, so he wants to duplicate that distance in the other direction. How would you do that? Well look, you can do this. Get out your compass, put the center at D, let DA be the radius, and draw a semicircle. Where it comes around and hits the line again, we'll call that point B, and sure enough, since you did this with a compass—and you're allowed to use compasses by Postulate 3—DA is the radius, DB is the radius, they're equal. So, you've constructed the segment DB equal to DA.

Euclid says, on the segment AB construct an equilateral triangle, ABC. Can he do that? Yes, remember Proposition I:1 showed you how, on a given segment, you can put up an equilateral triangle. If you didn't like this, if you challenged Euclid, how do you know you can this, he would refer you back to Proposition I:1. So, Proposition I:11 is certainly being built upon the shoulders of Proposition I:1, he builds the equilateral triangle there.

Then finally he draws the line CD, from C, the vertex of the equilateral triangle down to D, that original point, and he does that by Postulate 1; you can draw a line between any two points. So this is his picture, and now he's going to have to prove something about this particular construction. What he says is that triangle ADC, the one on the left, and triangle BDC, the one on the right, are congruent. So that's his claim and he has to justify that. Why is that? DA and DB are equal because they're radii, the two horizontal sides, they were radii. He knows that AC and BC are equal because you had constructed an equilateral triangle; those two sides are equal. Finally CD is equal to itself, it's the common side to the two triangles. So if you look at the left and right triangle, we have equal side horizontally, equal side oblique, equal vertical, side-side-side will give you congruence. So indeed, triangle ADC is congruent to triangle BDC, and that means that angle ADC must equal angle BDC, which I will call theta, so these common angles, these are equal, theta and theta on either side.

At this point, Euclid's just about done because now he says, that means CD is perpendicular to AB. Why is that? Remember what the definition was of perpendicularity. This is where he has to go back to his definition, and Definition 10 said that when a straight line standing on a straight line makes the adjacent angles equal to one another, each of these angles is right, and the straight line standing on the other is called a perpendicular to that one which it stands. It's just what he has got here. Theta on one side, theta on the other, this is a perpendicular, that's what he claimed he has constructed, a perpendicular to align at a point on it, Q.E.D., a very nice bit of Euclidean mathematics typifying his careful reasoning, his reliance on previous propositions, and his basic definitions and postulates.

From there, more theorems are coming, I'm just going to mention one last one, we'll jump ahead a few, Proposition I:20 is this. It says in any triangle, two sides taken together in any manner are greater than the remaining one. So this is an inequality. Something is greater than something. This is sometimes called the Triangle Inequality. The idea is, if I draw the picture here, you have a triangle, let's say the sides are A, B, and C. Any two of these added together is more than the third. So let's say we'll take A plus B, those two add up to be more than C. This would be the triangle inequality applied to those sides. As you might guess, he proves this using that Common Notion 5, the whole is greater than the part, and that figures in the proof. I'm not going to give the proof, it's actually rather sophisticated.

In even classical times, the Epicurean philosophers criticized this proof. They said, wait a minute, this proof isn't even necessary. Anybody knows the two sides of a triangle are longer than the third side. As a matter of fact, even an ass knows this, said the Epicureans. You know, why bother proving this. Here's what they meant. If you took a donkey and you put the donkey at one vertex of the triangle and you put the donkey's food at the other side, the donkey knows that it's much quicker to just go straight alongside C than to go around by A plus B. The donkey knows that those two sides are more than the third, so the donkey would go straight to the food, right?

So the Epicureans said, why even bother proving such a thing? It's just too obvious. Many centuries later a Greek commentator named Proclus who was a fan of Euclid's *Elements*, he wrote a treatise about Euclid's *Elements*, he took up Euclid's side of the argument, and Proclus said, no, no, we have to prove this. Euclid's very mission was to prove things, even if they seem self evident. Here's his words: Proclus said, yes, yes, granting that the theorem is evident to sense-perception, it is still not clear for scientific thought. Many things have this character; for example, that fire warms. This is clear to perception, but it is the task of science to find out how it warms.

So, maybe the triangle inequality is evident, even to an ass, but why is it true? That is Euclid's mission and that Euclid accomplished beautifully.

With this, we are deep into Book I of the *Elements*. We are going to stop this lecture here, but I will be back to pick up more results including the great Pythagorean Theorem, and it's converse, the theory of parallels, and lots of other things that come up later in this great masterpiece.

Euclid's *Elements*—Triangles and Polygons
Lecture 4

Thousands of years later at the end of the 18th century ... Carl Friedrich Gauss looks back into Euclid and finds something that Euclid had missed. When we get to that—many, many lectures down the road—we will be constructing regular polygons again. This is one of those things [where] "what's old is new again."

We continue our journey through Euclid with proposition I.26, proving the remaining congruence schemes: angle-side-angle and angle-angle-side. These give Euclid a full complement of congruence schemes, which constitute a critical feature of his geometry. In proposition I.27, Euclid introduces the concept of parallels for the first time. According to this proposition, if alternate interior angles are congruent, then the lines are parallel. In I.29, Euclid proves the converse—if two lines are parallel, the alternate interior angles are equal—using postulate 5 (the parallel postulate) for the first time. For the remainder of Book I, he uses this postulate in every proposition, except I.31. His proof of I.29 is done by contradiction.

A few propositions later, in I.32, Euclid uses I.29 to prove an important result: The angles of a triangle must sum to two right angles. As we saw earlier, Thales supposedly completed this proof long before Euclid. Proposition I.46 is also important; this shows how to construct a square on a finite straight line. As you recall, proposition I.1 showed how to construct an equilateral triangle on a line segment, but Euclid couldn't prove the same result for squares until he had established the propositions for parallel lines. Propositions I.47 and I.48 are the highpoints of Book I. According to I.47, in right-angle triangles, the square on the side subtending the right angle is equal to the squares on the sides containing the right angle. You might not recognize it, but that's the Pythagorean theorem. Euclid's graphic proof of this proposition is shown in his "windmill diagram." Proposition I.48 proves the converse: A triangle in which the square on one side is the sum of the squares on the other two sides is a right triangle.

Book III of the *Elements* deals with circles; the first proposition here shows how to find the center of a given circle. Book IV is about constructing regular polygons. Note that a polygon is regular if all the angles are equal and all the sides are equal. Proposition IV.11, for example, shows how to construct a regular pentagon; perhaps the most impressive proposition in Book IV is the last one, in which Euclid shows how to construct a regular pentadecagon, a 15-sided polygon.

If you've studied trigonometry, you know that it is based almost entirely on the Pythagorean theorem and similar triangles.

In Book VI, Euclid turns to similar figures, which are those that have the same shape but not necessarily the same size. In proposition VI.4, Euclid proves that in triangles with equal corresponding angles, the corresponding sides are proportional. If you've studied trigonometry, you know that it is based almost entirely on the Pythagorean theorem and similar triangles. Proposition VI.8 states: In a right triangle, the altitude to the hypotenuse creates two smaller right triangles, similar to each other and to the original triangle. We'll see this famous result again later in the course.

In the next lecture, we'll turn to Book VII, in which Euclid takes up a completely different topic: the phenomenal subject of number theory. ∎

Suggested Reading

Euclid, *Euclid's Elements*.

Heath, *A History of Greek Mathematics*.

Proclus, *A Commentary on the First Book of Euclid's Elements*.

Questions to Consider

1. Read and understand proposition I.47, Euclid's famous "windmill" proof of the Pythagorean theorem. This can be found, for instance, at http://aleph0.clarku.edu/~djoyce/java/elements/bookI/propI47.html. Euclid's

is not the simplest proof of the Pythagorean theorem, but it is surely one of the classic arguments in the history of mathematics.

2. If a right triangle (as shown) has base 5 feet and area 10 square feet, find its perimeter.

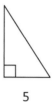

5

Euclid's *Elements*—Triangles and Polygons
Lecture 4—Transcript

In the previous lecture, I introduced Euclid's *Elements*. We saw his definitions, his postulates, his common notions, and took a tour through some of the results in the first half of Book I, such things as his side-angle-side congruent scheme, his side-side-side congruent scheme, and the triangle inequality.

What I want to do in this lecture is continue our journey through Euclid, to see more results about triangles and their extended family, the polygons, and see if we can get through to Book Six for the quick tour.

Proposition 26 of Book I are the remaining congruent schemes, the angle-side-angle and the angle-angle-side congruent schemes, the ASA and the AAS as we would call them. That gives Euclid his full complement of congruent schemes, ways to show triangles are congruent, and this is a critical feature of his geometry.

Then, in Proposition I.27 he changes radically because he is now, for the first time, going to introduce the concept of parallels. Remember he had defined parallels back at the beginning of Book I, but he had not mentioned them ever since, the first 26 propositions have nothing to do with parallels. It was as if he was holding them in the wings. But now, parallels are going to come to play an important role in his geometry.

The first place where this happens is the next proposition, Proposition I.27. Here's the diagram that comes with it. You see line *AB*, line *CD*, a line crossing through these transversal as it's called, and two angles alpha and beta, these are called the alternate interior angles.

Proposition I.27 says this: If alpha equals beta, then *AB* is parallel to *CD*. If the alternate interior angles are congruent, we would say then, the lines are parallel. So for the first time, parallelism is mentioned and Euclid proves this particular theorem. But it's not this theorem I want to focus on. It's I.29, two propositions later he addresses the converse. So, we have the same diagram but remember with converses we are going to interchange hypothesis and conclusion, so I.29 reads this way: If *AB* is parallel to *CD*, then alpha equals beta; then the alternate interior angles are the same. This is an important landmark of the development of Book I, because this is the first time Euclid uses Postulate 5.

Remember we saw the postulates, that fifth one was kind of wordy, kind of complicated, and in fact Euclid avoided using it up until this proposition. The other postulates, the common notions, he used them willy-nilly, but Postulate 5 he was sort of holding at bay until he couldn't hold it at bay any longer. He needed it here in I.29, and it turns out, once he used it he couldn't give it up, because for the whole rest of Book I he uses it in every subsequent proposition except Proposition 31. So, once it has hit the stage, it runs pretty much center stage all the way.

Let me show you how he proves I.29. He is going to use, as I said, the parallel postulate, Postulate 5, and let me just remind you what that says: When a straight line falling on two straight lines makes the interior angles on the same side less than two right angles, the two straight lines if produced indefinitely, meet on that side on which the angles are less than the two right angles; a real mouthful.

The picture, remember, was this, *EF* falling on lines *AB* and *CD*, and the postulate said: If alpha plus beta is less than two right angles, then *AB* and *CD* will meet to the right. That is going to get used in his proof of Proposition I.29. Here's how he does it, I.29, this is the theorem, remember, assume the lines are parallel, *AB* and *CD* are parallel, show that the alternate interior angles are equal. His plan of attack is an interesting one because he is going to prove this by contradiction.

We haven't seen one of those before in Euclid, but he was a master of proof by contradiction, the so-called indirect proof. How that works is you assume *AB* is parallel to *CD*, but suppose that alpha is not beta, suppose the conclusion isn't true. If the conclusion is supposed to be that alpha is beta, Euclid assumes it isn't, and from this, if he can derive a contradiction, then the rules of logic allow him to reject that assumption that alpha is different from beta and the conclusion would be, as he wishes, alpha equals beta.

So, he says look, if alpha is not equal to beta, either alpha is bigger or smaller, and it really doesn't matter which, the proof goes the same in either direction. So let me just assume that alpha is smaller than beta. That's why they're not equal. So, we assume alpha is less than beta. Next to beta there, along the line *CD* I'm going to put in angle gamma. If alpha is less than beta and you add gamma to both, then alpha plus gamma is certainly less than beta plus gamma. You've just increased both of these by gamma, the inequality remains the same. But beta plus gamma are the two angles around a straight line, and so they sum to two right angles. So now we have this scenario where you have line *AB*, you have line *CD*, you have this line cutting across, and alpha plus gamma, that is the two angles interior between the two lines on one side of the transversal, these sum to less than two right angles.

Postulate 5 guarantees that lines *AB* and *CD*, therefore must meet to the right. That's what the postulate says happens. So Euclid is justified in making that conclusion. But wait a minute, the lines *AB* and *CD* were assumed to be parallel, and by applying Postulate 5 here I've just concluded that they must meet. This is a contradiction. The one thing we know about parallels is, they never meet; and so we have reached a contradiction to the parallelism of *AB* and *CD* that allows me to refute the original supposition that Euclid made in this proof by contradiction. He started out by saying, suppose alpha is not beta, uh-uh, that is not true. Conclusion, alpha is beta and the indirect proof is complete. It's a very nice argument. It employs this proof by contradiction technique as well as the first appearance of Postulate 5. So now he knows if you have parallel lines cut by a third, the alternate interior angles are equal.

A few propositions later he proves a very important result using this: Proposition I.32, the angles of a triangle must sum to two right angles. Remember we mentioned that Thales supposedly did this long, long before Euclid, but here's the proof that we do have from Euclid's *Elements*. So he has got a triangle *ABC*, he has got to show that alpha plus beta plus gamma adds up to exactly two right angles. How do you do this? Through the Point *C* he says, Let's start by drawing a parallel to *AB*. So he constructs a parallel through *C*, let's call it *DE* to the line *AB* and he has shown in earlier works, in earlier propositions, how to do this with a compass and straightedge. So, let's just go with him and put in that parallel.

DE is parallel is *AB*, we have lines cutting through so I'm going to be able to employ Proposition I.29 just proved, and this Euclid does. If I look at the two parallels *AB* and *DE*, and I think of *CA* as being the transversal cutting through them, I then know that angle alpha down there has as its counterpart the equal angle *DCA*. That's going to be an alpha, I put it in. Likewise *DE* and *AB* are parallel, but now think of *BC* as being the transversal cutting across. Therefore angle beta will have as its equal counterpart, angle *BCE*, so that's going to be a beta up there, so I put that in.

So this is the way the diagram looks now. Now what I do is look at the line *DCE*, going straight across there. You see there's an alpha, there's a gamma, there's a beta around one side of the line. Well that means that alpha plus beta plus gamma up there around the point *C* must add up to two right angles, the sum of the angles on one side of a line, we would say, is 180 degrees, the Greeks would say two right angles.

Now the proof is done, particularly if you just forget about the alpha and beta up high there that I put in, let me remove them, I don't need them anymore. Away goes the alpha, away goes the beta, and look what I've got. I still conclude alpha plus beta plus gamma is two right angles, but lo and behold, those are the three angles of the original triangle *ABC*; just as I had hoped, the proof is done thanks to Proposition I.29, this important result about parallels. So, triangles' angles sum to two right angles, one of the most important results in Euclid's geometry.

He continues, I'm going to jump ahead to Proposition I.46, almost done, there's 48 propositions in Book I. This one is an important one. He shows how, on a finite straight line, to construct a square with compass and straight edge. So you give me a line *AB* and he wants to put up a square there, how to construct this, and he shows how to do it. Now this raises an issue. Remember the first proposition of Book I, the very first thing he did was construct an equilateral triangle on a segment, the sort of perfect triangle, three equal sides, three equal angles. That was how we started Book I.

Here he is doing the perfect quadrilateral, four equal sides, four equal angles, but it takes him all the way to Proposition I.46 to do this. Why didn't he do this earlier? Why didn't he do this back near where he did the equilateral triangle? The reason, of course, is that squares have parallel sides and you can't do this without knowing something about how parallels behave, and Euclid had put that off to the second half of Book I. Therefore, this wasn't an option until late in Book I. But he does it, he constructs a square on a given segment and that is going to be real important when the next proposition comes along. This is the biggie, Proposition I.47 and it's companion Proposition I.48 form the climax of Book I; I.47 particularly is important. In right-angled triangles it says, the square on the side subtending the right angle is equal to the squares on the sides containing the right angle.

Well, you might not recognize that, but that's the Pythagorean Theorem. In right triangles, the square on the side subtending the right angle means the square on the hypotenuse, this equals the squares on the sides containing the right angle.

Here's a page from an Arabic manuscript that contains a proof of the Pythagorean Theorem, Proposition I.47, you can maybe see it there in the diagram, I'm going to put up a diagram in a minute, a little larger, but the name of this picture has come down to us as Euclid's "windmill" diagram. It kind of looks like a windmill, the way this proof works because he takes the right triangle, puts squares on the legs, square on the hypotenuse, it sorts of looks like the blades of a windmill, so the common term for this is the "windmill" proof. I'm not going to give the proof from Euclid I.47, it's a little bit complicated and we already have proved the Pythagorean Theorem a few lectures ago, but let me just suggest the logical strategy that he adopts to do this great result.

He takes his right triangle, *ABC*, and he puts a square on the hypotenuse *AB*. Can he do this? Yes, remember I.46, the previous result, he showed you how, given a segment, to construct a square, so he's ready for you, he puts the square on the hypotenuse *AB* and while he's at it, he puts squares on the other two legs, again citing Proposition I.46 which he has dutifully proved. The squares I've put on the two legs, the square on *AC* and *BC*, I have colored in here blue and green. That will be important for the strategy of his proof. What Euclid does now is through the point *C* he constructs a parallel to the left-hand side of that square on the hypotenuse. Another way of thinking of it would be this, drop a perpendicular from *C* to line *AB* and just keep going.

It amounts to the same thing. So I've drawn that line in, this line from C parallel to the left-hand side of the square on the hypotenuse.

Then, using some very interesting geometry, he shows that the area of the blue square up there on side AC is exactly the same as the area of the rectangle formed to the left of this line he just drew downward. In other words, the blue area of the square is the blue area of that rectangle. I'm not going to go through the argument, it involves some congruent triangles, some interesting propositions, but he shows the blue area equals the blue area. In an exactly analogous fashion, he shows the green area of the square on BC equals the green area of the rectangle below, voila.

Now, says Euclid, the sum of the squares on the legs, that is the blue square up there plus the green square up there, is going to equal the blue rectangle below plus the green rectangle below, but look at the picture. The blue area plus the green area below add up to the square on the hypotenuse, and by that subdivision proof, he establishes that blue square plus green square equals square on the hypotenuse. It is a very nice argument with this windmill shape formed here.

That's Proposition I.47, maybe the most famous result in Euclid's *Elements*. But he's not quite done with Book I; being a good logician, he wants to tidy things up and get the converse proved, and that's Proposition I.48.

So here comes I.48. It says this: Suppose you have a triangle where the square on one side is the sum of the squares on the other two, C squared equals A squared plus B squared. Does it follow that that's a right triangle? That's the converse. Well, that's interesting. The answer is yes, it does. This proof goes both ways, and Proposition I.48 is the proof of the converse of the Pythagorean Theorem. So let me show you how he proves this.

He begins by constructing segment CD perpendicular to AC. So if along the line AC I want to put a perpendicular outward from the point C, well that's what we saw him do in Proposition I.11 in a previous lecture. That's exactly what he needed, he cites that and builds up a perpendicular CD. Furthermore, he makes it of length A so the length of CD is exactly the same as the length BC. You could do this by putting the perpendicular outward from C and then using your compass to make a circle whose radius is *A*, it would cut through *B* and it would cut around through *D* and you would have achieved this result.

Next he says, draw *AD*, this line, and call its length *d*. So there's the diagram that he's going to be needing to prove the converse of the Pythagorean Theorem. Interestingly, he's going to use the Pythagorean Theorem in this proof. He cannot apply the Pythagorean Theorem to the triangle on the right, triangle *ABC*, we don't know that's a right triangle. In fact, that's the whole object of this theorem; but the triangle on the left, triangle *ACD* is a right triangle because you've constructed that perpendicular there, so I can certainly apply the Pythagorean Theorem to that one. Proposition I.47, the Pythagorean Theorem is applied to right triangle *ACD* to conclude that *d* squared, the hypotenuse squared is *a* squared plus *b* squared, the sum of the squares on the legs, so that's legit.

Now he says, what was my hypothesis, what was my assumption in Proposition I.48, it said that *c* squared is *a* squared plus *b* squared. Remember, that's how we began this. Look at where I am, *d* squared is *a* squared plus *b* squared, *c* squared is *a* squared plus *b* squared by assumption, so *d* squared equals *c* squared using the theorem's hypothesis. Well, if *d* squared equals *c* squared, then *d* equals *c*. This would follow. We would say, just take the square root of each of these and you get that result.

After this little argument employing the Pythagorean Theorem as well as the assumptions in Proposition I.48, he has concluded that that length *d* is the same as the length *c*. Let me draw the picture again, there's my picture. Watch this, the length *d*, I'm going to change to what it equals to *c*. So there I've made the change and now this is what the picture looks like.

At this point Euclid says that those two triangles left and right are congruent. Triangle *ABC* and triangle *ADC* are congruent, why, well look. They both have a side *C*, they both have a side *A*, they share a side *B*, side-side-side. That congruent scheme comes into play, these triangles are congruent which means they are equal in all respects, which means angle *BCA*, the angle of my original triangle equals its counterpart, angle *DCA*, but angle *DCA* was constructed to be a right angle, remember. Hence, angle *BCA* is a right angle and that means triangle *ABC* with which we began is a right triangle. That's what the converse of the Pythagorean Theorem called for, Euclid did it. It's a very nice argument, a good way to end Book I.

We've got that great first book of Euclid out of the way. Let me jump to Book III, which is about circles. This is his book about circles. Now you say, we already did circles, we have drawn circles all the time and it's true, he has employed Postulate 3 to draw circles, but the circles were in the service of theorems about triangles. Now he wants to just study circles all by themselves, and let me just mention one result from here, the first proposition, Proposition III.1, he shows you how to find the center of a given circle.

Now you might say, why would I need to do this? If I have to make a circle, there is a center, I put my compass there, I get my radius, I draw it, I know where the center is. Why must I find the center of the circle? I think of it this way, suppose you did that, you drew your circle and then you left the room, the circle was still on the board, you have erased the center and somebody else comes in. Can they recover that center, can they figure out where it was, and yes, you can, and Euclid gives a very nice proof how that is done. There is this result and many other results that show up about circles in Book III.

Book IV is about constructing regular polygons. A polygon, of course, is a figure with many sides, as many straight lines as sides, but a regular polygon, we need to kind of remember what that means. A polygon is regular if all of its angles have the same size, the same measure, and all of its sides have the same length. So it is a perfectly symmetric, beautiful polygon, it is regular. All of the angles are equal, all the sides are equal, this would have appealed very much to the Greek sense of aesthetics, this kind of symmetry, and we've already seen a few of these. An equilateral triangle is an example of a regular three-sided polygon, and Euclid shows us how to make this remember in Proposition I.1. He had constructed an equilateral triangle in a segment.

A square is a regular quadrilateral, regular four-sided figure, four equal sides, four equal angles, and Euclid had showed us how to do this in Proposition I.46. So we have already seen a couple of regular polygons, but in Book IV he wants to expand the repertoire.

Proposition 11 of Book IV, he shows you how to construct a regular pentagon. He starts with a circle in which, by a lot of sophisticated constructions, he figures out how to inscribe a regular pentagon, five equal sides, five equal angles. This is not trivial. If you think it's easy, try it. This takes a bit of doing to try to create a regular pentagon, but he does with the Euclidian tools, compass and straight edge. So that's pretty impressive, but the most impressive result, I think, is the last proposition of Book IV where he shows you how to construct a regular pentadecagon. Penta is five, deca is ten, a pentadecagon is a 15-sided polygon. He can show you how with a compass and straightedge you can construct a regular polygon with 15 equal sides, 15 equal angles.

Let me show you how he does this, let me just give you a sense of this, because this is quite the climax of Book IV. He starts with a circle and picks a point A to be a vertex. Starting at A using that as a vertex, he shows you how to inscribe within the circle an equilateral triangle. So there it is, that's already something he could do. He could do triangles, he puts in a triangle.

Let me let C be the point on the circle where that first side of the triangle ends, so AC is the side of the equilateral triangle. Now he says, also starting at A, inscribe within the circle a regular pentagon, and remember he could do these too. He could do pentagons, he already had done that, so there is a perfect 5-sided figure, and let me let B be the place where the second vertex of the pentagon is, down around the corner there. So then we have these two points, B and C, singled out along the circle. What Euclid says is, the arc from A around to B is exactly two-fifths of the circumference, two-fifths of the way around because it is two of those five equal sides.

Meanwhile, the arc from A to C is exactly one-third of the circumference because that's along one side of the equilateral triangle, that's exactly one-third of the way around. What he wants to know is: How much of the way around is that little arc from B to C? That arc from B to C is exactly two-fifths which is the arc from A to B minus one-third, the arc from A to C. So the little arc from B to C is exactly two-fifths minus one-third of the circumference, exactly. But what is two-fifths minus one-third? You get your common denominator of 15 and that will be six-fifteenths minus fivefifteenths, one-fifteenth, the arc from B to C is exactly one-fifteenth of the way around.

What you do is, you take out your compass, you figure out that length from B to C, you could draw a little straight segment there if you wanted, use your compass, copy that and you can copy it exactly fifteen times and you will go right around and come back to where you started from, you will have made a regular pentadecagon. It was a brilliant piece of reasoning.

Believe it or not, this whole question of constructing regular polygons will come back thousands of years later at the end of the 18th Century in the work of Carl Friedrich Gauss who looks back into Euclid and finds something that Euclid had missed. When we get to that, many, many lectures down the road, we will be constructing regular polygons again. This is one of those things that's "what's old is new again."

Let me jump to Book VI where Euclid is talking about similar figures. You'll remember similar figures are those with the same shape but not necessarily the same size. You think of them as enlargements of one another. I have drawn these two triangles here and they seem to have the same shape, but they are not the same size, so these are going to be similar.

What Euclid proves in Proposition IV of Book VI is that if triangles have their corresponding angles equal, then their corresponding sides are proportional. They will be similar, they will have their corresponding sides proportional.

All you need to do for similar triangles is show that the angles are the same, one to one, this angle equals that, this equals that, and the sides will be proportional. So in my picture, the angles indeed are the same of these two triangles, and hence A is to B, that ratio, is the same as C to D. The proportionality means that they are not the same lengths, A and B are not the same lengths as C and D, but their ratios are the same. That is why similarity is so important. If you've studies trigonometry you know that trigonometry is based almost entirely on the Pythagorean Theorem and similar triangles.

Euclid has now gotten the concept of similarity out there, he has proved that in triangles, to show they are similar, all you need to do is show that their angles are the same.

Then he proves one result which I'm going to look at that is an important one involving similar triangles and is something we are going to need later in the course, so this would be a good time to take a look at it. It is Proposition 8 of Book VI. There Euclid says: In a right triangle, so a particular kind of triangle, the altitude to the hypotenuse creates two smaller right triangles, each similar to one another, and as a matter of fact, each similar to the original triangle you started with.

Let me show you how this works. We have a big right triangle, *ABC*, so there it is. Let's suppose that the angle at *A* is called alpha, the angle at *C* is called beta. What I'm going to do is draw the perpendicular from *B* downward hitting the hypotenuse at *D*. I claim that in so doing, the triangle to the left of that vertical line, triangle *ADB* is similar to the one to the right of the vertical line, triangle *BDC*. That is pretty easy to do by just tracking down the angles and showing that the two triangles have exactly the same angles in them.

One thing we notice is, from the original right triangle, triangle *ABC*, we know that the angle alpha plus the angle beta plus that right angle up at the top has to be two right angles. Any triangles have its three angles add up to two right angles. If I subtract a right angle from each, I know that alpha plus beta therefore must be one right angle, a famous result about right triangles.

Look at triangle *ADB*, the triangle to the left of the dotted line, it has a right angle in it, it has an alpha in it, because alpha plus beta is a right angle, the other angle there has to be beta. I can fill in angle *ABD* up there as a beta. While we're at it, look at angle *DBC*, the one right next to the beta I just filled in. That is the right angle minus beta will fill in that angle *DBC*, that must be an alpha. This is how the picture has to break down for right triangles, if the two angles are alpha beta, so are those two angles up at the top. That means the triangles are similar because they have the same angles. Each of the triangles, *ADB*, *BDC* and, for that matter, the original triangle, have an alpha angle, a beta angle, and a right angle so they are similar.

I can use that in the following sense, if I have my right triangle here, I have drawn the perpendicular downward, let's suppose it splits the hypotenuse, this perpendicular splits the hypotenuse into two chunks, two pieces, one of which is of length x and the other of which is of length y. So here comes the altitude down from the right angle to the hypotenuse splitting the hypotenuse into x and y, and suppose the altitude has height h. Then I have just shown that the two triangles, the blue one on the right, the green one on the left, are similar. They are not the same size but they are the same shape. By the similarity, a proportionality follows that is going to be real important, namely in the green triangle the long leg x is to the short leg h in exactly the same ratio as in the blue triangle, the long leg h is to the short leg y.

If you then cross multiply this equation, x over h equals h over y, cross multiply you would get that h squared is xy, a famous result about perpendiculars to the hypotenuse in a right triangle. We will see this again later in the course.

With that, I am done with Book VI. Interestingly, in the next book, Book VII, Euclid is going to change entirely. We are going to go off in a completely different direction. It is really quite amazing and quite beautiful as he takes up the phenomenal subject of number theory. Stay tuned.

Number Theory in Euclid
Lecture 5

> People that like math will look at this proof of the infinitude of primes, done in this sort of indirect fashion, and be just enthralled. They are the people that should become mathematicians.

Many people are unaware that Euclid's *Elements* deals with number theory in addition to geometry. Number theory is the study of the properties of whole numbers. It may seem as if these are the simplest sort of mathematical entities, but in fact, number theory is one of the most difficult mathematical subjects.

Euclid begins, again, with definitions, the first of which reads as follows: "A unit is that by virtue of which each of the things that exists is called 1." In other words, the number 1 is a unit. According to the second definition, a number is a multitude composed of units.

In proposition 2 of Book VII, Euclid proves the Euclidean algorithm, which is a method for finding the greatest common divisor of two numbers. He also defines a prime number, which is a whole number greater than 1 that is divisible only by 1 and itself, and a composite number, that is, a whole number greater than 1 that is divisible by some number strictly between 1 and itself (e.g., 15 is composite because it is 3 × 5, and 3 is intermediate between 1 and 15).

Using the definitions of prime and composite, Euclid proves proposition VII.31: Any composite number has a prime divisor. The number 30, for example, is the product of 10, which is not a prime number, and 3, which is prime. The number 120 is the product of 10 and 12, neither of which is prime, but 10 can be broken down into 2 × 5, and 2 is prime. Euclid's proof of this proposition for any composite number is a classic result of number theory. His argument showed that any given number can be broken down into a finite chain of intermediate divisors in which each divisor is greater than 1 and smaller than its predecessor. The chain stops when we reach a prime divisor.

In Book IX, Euclid proves that there are infinitely many primes. An easy way to prove infinitude of some entity is to identify a pattern and just spin out results, as we see with the example of the squbes. But this method wasn't available to Euclid because there is no pattern for generating primes. In proposition 20 of Book IX, Euclid stated his theorem about primes as follows: "Prime numbers are more than any assigned multitude of prime numbers." In other words, no finite collection of primes contains all the prime numbers. If we start with a finite collection of primes—a, b, c, d, e—and we define n as the product of these primes + 1, our results will fall into one of two cases: (1) n itself may be a new prime number or (2) n will be composite and will have a prime divisor (p) that is not among the original primes.

Any given number can be broken down into a finite chain of intermediate divisors in which each divisor is greater than 1 and smaller than its predecessor. The chain stops when we reach a prime divisor.

Book X of the *Elements* is about something called quadratic surds; Books XI and XII venture into the realm of solid geometry; and in Book XIII, Euclid proves that there are only five regular solids (also called Platonic solids): the tetrahedron, cube, octahedron, dodecahedron, and icosahedron. With that proof, the *Elements* comes to an end. Ivor Thomas, a math historian, likened Euclid's work to the perfection of the Parthenon. The *Elements* is a great legacy of the Greeks' affinity for beauty and symmetry. ∎

Suggested Reading

Euclid, *Euclid's Elements*.

Hardy, *A Mathematician's Apology*.

Heath, *A History of Greek Mathematics*.

Ore, *Number Theory and Its History*.

Weil, *Number Theory*.

1. (a) True or false: There are infinitely many *odd* **prime** numbers.

(b) True or false: There are infinitely many *odd* **composite** numbers.

2. Revisit Euclid's proposition IX.20, in which he proved that no finite collection of primes can contain *all* the primes. Start with primes $a = 2$, $b = 5$, $c = 7$, and $d = 11$ and form the number $N = (a \cdot b \cdot c \cdot d) + 1$. Show that N is not prime, but show as well that its prime divisors are "new" in the sense that they are not 2, 5, 7, or 11. This is exactly what Euclid proved in case 2 of his wonderful argument.

Number Theory in Euclid
Lecture 5—Transcript

In our previous two lectures, we have surveyed the first six books of Euclid's *Elements* where he does his plane geometry. Now there is a big change. In Books VII, VIII, and IX, Euclid investigates number theory. A lot of people that have heard of Euclid's *Elements* are not aware that it is a great number theoretic text, as well as being a great geometry text. Number theory is the study of the properties of the whole numbers, the counting numbers like 5, and 11, and 28. It seems like the simplest sort of mathematical entities. As a matter of fact, number theory is one of the hardest subjects in all of mathematics, and its first great early incarnation is in Euclid's owns.

Because he is going to a whole different subject, he has got to give new definitions. He's got to start off with number theoretic definitions. Let me just share a few of these with you. His first is this. He says, "A unit is that by virtue of which each of the things that exists is called 1." I think most of us agree that is pretty confusing. What in the world does that mean? For our purposes, let's just say 1, the number 1, is a unit. Actually, I shouldn't call it the number 1. To Euclid, 1 isn't a number, it is a unit. Definition 2 says, a number is a multitude composed of units. So for Euclid, the first number, the smallest number is 2; 1 is something else, a unit.

In Proposition 2 of VII, Euclid proves the Euclidian algorithm. He introduces and proves this is a method for finding the greatest common divisor of two numbers, the gcd as it is called, the greatest number that divides into both. So, if I wanted the gcd of 12 and 20, what I'm asking is what is the biggest number that commonly divides into both 12 and 20? That is easy. That's 4. You don't need any special algorithm for that. But if I said what is the gcd of 110,207 and 2,295,820, you are lost. How in the world do you do this? How do you find the greatest common divisor of those two giant numbers? Well, Euclid shows you. It's called the Euclidian algorithm. It's central to all number theoretic books even to this day. If you pick up any of modern number theory books, right away they are going to be proving and using Euclid's algorithm from so long ago. By the way, the answer is the gcd is 191, and I did that with Euclid's algorithm.

He then defines a prime number. A prime number, he says, is a whole number greater than 1 that is divisible only by 1 and itself. A prime has no other divisors than 1 and itself. So, for example, 2 is a prime number. 17 is a prime. The only things that divide into 17 evenly, the only whole numbers, are 1 and 17, there are no others. So in 65,537, that's a prime, but that's a little harder to show that it can have no other divisors but 1 and itself.

Then he defines a composite number, a counterpart of the prime. This is a whole number greater than 1 that is divisible by some number strictly between 1 and itself. So a composite number has an intermediate size divisor, not just 1, the smallest divisor possible and itself, which of course is the greatest. There is some intermediate size one. Another way of thinking of this, a little more symbolic, is this. N is a composite number if you can write N as a product of $a \times b$, where the a is bigger than 1 but less than N. There is that intermediate size divisor a. Such a number is called composite. Examples, well, 15 is composite because it is 3×5 and 3 is intermediate between 1 and 15. 49 is composite, 7×7. The numbers could be the same as long is there is an intermediate size divisor.

So, he has got the big terms of prime and composite on the table. Then, Euclid proves in Proposition 31 of Book VII, this very important idea. Any composite number has a prime divisor. You can see this at work in an example like 30, right, 30 you can write as 3×10. The 10 isn't prime, of course, but the 3 is, so 3 is a prime divisor of the composite number 30. Suppose I took 120, and that's a composite, and I can write it as 10 times 12. Break it down into two pieces. It's composite. But notice, neither of those is a prime, neither 10 nor 12. So I have to break the 10 down; 10 is 2×5. Now what I say is, 2 divides evenly into 10; 10 divides evenly into 120, so the 2 divides evenly into the 120, and the 2, of course, is a prime. So, there is a way to get a prime divisor of 120.

What Euclid says is, any composite number has a prime divisor, not just 120 and 30, any does; and I want to show you his proof. This is a classic result from number theory. The proof goes like this. It is Proposition 31 of Book VII. Any composite has a prime divisor. He said, all right, let's let N be a composite number. What does that mean? Well, we have a definition for composite. It means you can write in as $a \times b$, a product of two pieces, where the a is intermediate in size between 1 and N. So, a is bigger than 1, less than N. So that much I know. So, N is so expressed, $a \times b$. Euclid says, if a is a prime, then we're done. What am I trying to show? Any number N has a prime divisor. If a is a prime, it is certainly a divisor. You got it. So then we're finished. But maybe it isn't a prime. There is no guarantee. If it isn't a prime, then a is a composite. So now Euclid has to go see what that means. Well, back to the definition. If a is composite, a can be written as a product of $c \times d$, two pieces where the c is less than a and bigger than 1. But remember, the a itself was less than N. So now I've got this chain; 1 is less than c, is less than a, is less than N. If c is prime, we're done. Why is that? Well, c would be prime, c would divide evenly into a because a is $c \times d$. A divides evenly into N because N is a times b. And hence, c divides evenly into N. It's just like in my numerical example, the 2 divided evenly into the 10, the 10 divided evenly into the 120, so the 2 does. So, if c were prime, c would be a prime divisor of the composite number N.

Great, but it doesn't have to be. So, maybe c is composite. Now what? Well, back to the definition. If c is composite, c is equal to $e \times f$, where e is intermediate between 1 and c. So, 1 is less than e is less than c. But we said previously that the c is less than a, and we said before that the a is less than N. So we're getting this chain being created here. As you continue this, you're going to have, here is my chain, 1 is less than whatever is coming next. Eventually, you get less than e, less than c, less than a, less than N. Notice that each new divisor I create in this fashion is greater than 1, but smaller than its predecessor.

Euclid at this point says, we cannot fit infinitely many such whole numbers in the range between 1 and N. We cannot fit in infinitely many numbers. Of course not, between 1 and 120, if you've got to put whole numbers, then they've got to be less than 120 and more than 1, there is only 118 slots. So after 118 steps at the most, you'd have to stop because you cannot fit infinitely many numbers in here. When I come to the split in the road, at each case it is either prime or composite. It is either prime or composite. I can't go on forever. The process of finding these new divisors must stop, and Euclid asked, what stops it? When does the process stop? It stops when we get a prime. So, as I move down the chain, I go from N to a to c to e, and I might go for a long time, but eventually, I've got stop. What will stop this process so I do not go beyond is I will have a prime divisor of each of the numbers up the chain, including N. Any number, any composite number N must have a prime divisor. That's why. That's the proof in Euclid. If you look in a modern number theory book, I promise you, that's the proof that's still there.

So, let me look at an example here, just to see this working. Suppose I looked at the hideous number 7,844,067, believe it or not, that's composite. I'm looking for a prime divisor. So what would I do? I'd break it up into two pieces. It turns out it breaks into 1,617 and 4,851. And the 1,617 is intermediate between 1 and the 7,844,067, but the 1,617 isn't prime. It breaks up into 21×77. The 21 is bigger than 1 but less than the 1,617, which in turn was less than the number we started with. The 21 can further be broken up into 3×7. Now I get the chain, 1 is less than 3, is less than 21, is less than 1,617, is less than 7,844,067; but I've reached a point where the process stops, 3 can't be further broken down. Why not? It's a prime. What Euclid's argument showed was this always has to happen with any composite as you come down the chain, the process stops. You've got the prime divisor. That's what he claimed. Any composite number has a prime divisor. 7,844,067 has the prime divisor of 3. So that's a major result.

Even bigger is in Book IX, when Euclid proves there are infinitely many primes. This is one of the greatest proofs in all of mathematics and we'll see it in a minute, but let me just alert you or preview what's coming. He's going to have to prove that there is infinitely many primes. You never run out of these things. Let me contrast what he is going to do, which we'll see in a minute, with a proof that goes much more easily to show there is infinitely many of something. For this purpose I'm going to introduce the notion of a *sqube*, which isn't a serious notion, but it will serve my purposes. So, here is my definition. A whole number is a sqube if it is simultaneously a perfect square, like 25, and a perfect cube like 8; $2 \times 2 \times 2$ is 8. So, a number that is both a perfect square and a perfect cube will be called a sqube. Bear with me on this. Here is an example, 1, pretty trivial right, 1 is equal to 1×1, so it's a square, and it's equal to $1 \times 1 \times 1$. So it's a cube; 65, that's a sqube, because it's 8×8 if it's a square, and it's $4 \times 4 \times 4$, that's a cube. And 729, that might not be so obvious, but you can try it out. It's 27×27, so it's 27 squared. And it is $9 \times 9 \times 9$; 9 cubed. So those are some examples of these squbes.

Suppose I was trying to prove there is infinitely many of these squbes. Remember what we're ultimately going to show is how Euclid shows there are infinitely many primes, but I'm working with squbes. Here is what I observed in order to do that. For any whole number n, n^6, $n \times n \times n \times n \times n \times n$, n is a sqube, n times itself six times. Why? Well look, n^6 could be written as something squared. What would have to be squared to get n^6; n^3 would have to be squared; n^3 times n^3, there is three ns × three ns, it's six ns altogether. On the other hand, n^6 is also something cubed. What is a cube? It is n^2 cubed; $n^2 \times n^2 \times n^2$, one, two, three, four, five, six there. So, any sixth power is a sqube because it's a square and it's a sqube. Well, with that background, I can now easily prove the theorem that there are infinitely many squbes. Why? Here is the proof. Let n be 1; 1^6 is a sqube because any six power is. Well, 1^6 is 1, sure; 2^6, which is 64, is a sqube; 3^6, which is 729, is a sqube; 4^6, 5^6, 6^6, they are all going to be squbes, and there are infinitely many of these, so voila, there are infinitely many squbes.

Why am I doing all this? Well, just to show you, that's the easy way to prove there is infinitely many of something. You just get a list, get a pattern, and just spin them out one after the other. Euclid doesn't do this. In his proof of Proposition 20 in Book IX, he doesn't go about it this way because he didn't have a pattern that generated primes, nor has anyone else found one. The proof that there is infinitely many primes is considerably more sophisticated, considerably more subtle, and it has to be, and yet Euclid did it back in Book IX. His statement was this. He didn't actually say there are infinitely many primes. He said, "Prime numbers are more than any assigned multitude of prime numbers." It sounds a little bit confusing, but here is what he means. No finite collection of primes, that is my assigned multitude, no finite collection of primes contains all of them. So if you give me 10,000 primes, they are not all there. If you give me 10 million primes, they are not all there. We can see that that means there are infinitely many primes. So, this is often called "Euclid's proof that there is infinitely many primes." It is actually his proof that any finite batch of primes can be augmented. You can find more primes than you already have.

Let me show you how he does this, grape theorem. He says, let's take a finite collection of primes. Let me call them a, b, c, d, up to e. Now there might be 15 there; there might be 15 million there. I don't know. It's a finite batch, and they are all primes. Remember now, his object is to show that there is a prime we missed; that batch does not contain them all. There are some more primes somewhere. He's going to have to show where they are, but he doesn't have a nice, easy pattern like the squbes. Here is what he says, I'm going to define a number N as follows: multiply all these primes that you've got, $a \times b \times c$, all the way up to e. There are finitely many of them. Maybe there are a thousand of them, so this is going to be a gigantic number if you multiply them all together, but you can do that because there are only finitely many. Then, add 1. So, N is the product of all the primes that you've got on the table, $a \times b \times c$, up to e, plus 1.

Two cases appear. Case 1, Euclid says suppose N itself is a prime, this new number you've created. Maybe it's a prime, maybe it is. If so, it's certainly a new prime that isn't in the original batch because look at it, it's way bigger than a or b or c, up to e individually. Remember, you multiplied all those original guys together. So you get a much bigger number, and then just for good measure, you add 1. So N is way bigger than a and b and c and e. So, if it's prime, it's a new prime that wasn't on the list, and you've just augmented your supply. So that's Case 1.

Case 2 is a little trickier. Maybe N isn't prime. It doesn't have to be. What if it isn't? Then this number N, $a \times b \times c$ up to e, plus 1 is composite. I still have to find a new prime not on the original list. Where do you find it? Euclid now says if N is composite, it has a prime divisor by the proposition I proved a minute ago, Proposition 31 of Book IV. Any composite has a prime divisor if N is composite, sure enough, it must have a prime divisor, which I'm going to call p for prime. So, I cite the earlier result, Euclid yet again building on his previous results. This number N, if composite, has a prime divisor p. Euclid says p is not one of the original primes. Remember, we're trying to augment the batch of primes from a to e. He says p is a new one. It's not there. It can't be. Why not; how about this? Suppose it were. Suppose this number p that you've just created as a prime divisor of N, suppose it were one of the originals, a or b or c, somewhere in that batch. If it were, then when I multiplied all the primes together, $a \times b \times c$, somewhere along the line I hit p as one of those primes, up to e. In that product, p would obviously divide evenly into that product because it's actually one of the factors. It's one of the terms. So, if p were in the original list, the prime p would divide evenly into that prime, $a \times b$

$\times c$ up to e. But, I just said p was also known to be a prime divisor of N. That's where it came from. It was a prime divisor of N. So, p would divide evenly into N, which is the product of all the original primes, $a \times b \times c$, up to e, plus 1. Dramatic conclusion is coming. If p divides evenly into N, and p divides evenly into the product from a to b to C, up to e, it divides evenly into the difference of the two. If a number divides into each of two numbers, it divides into the difference; 100 is divisible by 5; 85 is divisible 5; and the difference between them, 15, is divisible by 5. You can make a dollar out of nickels, 100 cents out of nickels. You can make 85 cents out nickels; and you make the difference, 15 cents, out of nickels, 5 divides into those numbers. So, if p divides evenly into N, and it divides evenly into $a \times b \times$ C, up to e, it divides evenly into the difference between them; but what is the difference between N and the product of all of the original primes? It's 1. Remember, N was just one more than that product. So p would have to divide evenly into 1. But wait a minute, p is a prime. It can't divide evenly into 1 because the smallest prime is 2. Remember, the first number you hit is 2; 1 is a unit, 1 is not a prime. A prime can't be 1. The first prime you hit is 2. So, the prime p cannot divide into 1. Therefore, p in Case 2 is a different prime than what was on the original list. It's a new prime. We have augmented the supply. What Euclid then says is, in either case, he has found the prime not among the original batch. The original batch does not exhaust them all. No finite collection of primes contains all the primes. There is infinitely many and that's his proof.

You can illustrate this. You can actually see this proof working here. So, let me do a numerical example to show this in action. Suppose our original batch of primes, our original collection, what he called our finite multitude, were 2, 3, and 5. Those are prime. What would we do? We'd create this number N, which is $2 \times 3 \times 5$. You'd multiply them all together and add 1. Well, that's 31; that's a prime already. It's way bigger than the 2 and the 3 and the 5. That's what Case 1 was about. You've created a new prime, 31, not among the originals, 2, 3, and 5, as in Case 1, because their product plus 1 is a prime. On the other hand, what if the original primes, the original finite multitude was 3, 5, and 7? Multiply those together and add 1 to create your number N, $3 \times 5 \times 7$ plus 1 is 106. Well, that's not a prime. So now we're in Case 2; but 106 has a prime divisor, 2 is a prime that divides evenly into 106, and it's not among the originals. The originals were 3, 5, and 7 in this Example B. The prime I created that divided into 106 is not among these. It can't be. That's what Euclid proves in Case 2. So, either way, whether this number N is prime or not, he shows you where to get a new prime, not in your original list. That's the proof of the infinitude of primes.

G. H. Hardy, a 20th Century mathematician of great note, looked at this argument from Euclid and said, "His proof of the infinitude of primes is as fresh and significant as when it was discovered — 2,000 years have not written a wrinkle upon [it]." This proof to Hardy is just as good as it was ever was. I would have to agree. Sometimes people use this proof as a kind of litmus test for someone's mathematical sense of aesthetics. If you looked at this proof and you say, "Nah, it doesn't interest me," then maybe you're just not cut out for math. But the people that like math will look at this proof of the infinitude of primes, done in this sort of indirect fashion, and be just enthralled. They are the people that should become mathematicians. I'm always struck how different this is from that sqube proof, where it was so easy to find the infinitude of squbes. You didn't have that option. Euclid wasn't blessed like that, but he did it anyway using his great powers of logic.

Well, that is in Book IX. Remember VII, VIII, and IX were his books on number theory. Then, as he finishes up through the remaining books, he changes gears again. Book X is about something called quadratic surds. These are rather complicated expressions, such as the square root of 2 plus the square root of 3, square roots embedded within square roots. He goes to a great deal of trouble to prove lots of theorems about these. Books XI and XII he ventures into the realm of solid geometry, 3-dimensional geometry, and proves things about solid figures. And then in Book XIII he addresses the regular solids, as they are called. A regular solid is something like a cube. It is a solid body, so it's 3-dimensional. Its faces are themselves regular polygons, as he addressed in Book IV, all the space angles are equal. So a cube is an example of a regular solid. It's a beautifully symmetric 3-dimensional body. What Euclid does in Book XIII is address the regular solids and proves that there are only five of these. Only five of these can ever exist. The five regular solids, these are sometimes called the "Platonic" solids because Plato had discussed these.

The Platonic solids are these: There is the tetrahedron, as it is called, four equilateral triangles as the faces in this triangular pyramid. There is the cube, six faces, each of which is a square, a regular 4-sided figure. There is the octahedron, eight faces, each of which is an equilateral triangle, perfectly symmetrical. There is the dodecahedron, 12 faces, each of which is a regular pentagon; and there is the icosahedron, 20 faces, each of which is an equilateral triangle. These are the regular solids. What Euclid proves in Proposition 18 of Book XIII, which is the last proposition in Euclid's *Elements*, he proves that there can only be these five, no other regular solids can exist. These bodies which would so much appeal to the Greek sense of symmetry and beauty are limited in number to these five. So with that, Euclid's *Elements* comes to an end.

I hope you've enjoyed this three-lecture survey of this work. I'm going to end with a couple of testimonials to Euclid's *Elements*. One of them is from the math historian Ivor Thomas. It's a little bit long, but let me read this to you. Thomas said this:

> [A] feature which cannot fail to impress a modern mathematician is the perfection of form in the work of the great Greek geometers. This perfection …, which is another expression of the same genius that gave us the Parthenon and the plays of Sophocles, is found equally in the proof of individual propositions and in the ordering of those separate propositions into books; it reaches its height, perhaps, in the *Elements* of Euclid.

I would certainly agree with this except for the "perhaps" part. I think it certainly reaches this height. I hope I've managed to show you the proof of the individual Propositions, or at least a few of them, and how these were so beautifully ordered in Euclid's *Elements*. So, Thomas is saying that Euclid's *Elements* is like the Parthenon in theorems, if you will. I might twist that around and say the Parthenon is like Euclid's *Elements* in stone. In any event, it's a great legacy of Greek thinking.

I'll end with a quotation from a great Islamic mathematician, Djamal Ibn Al-Qiftī, who looked at Euclid's *Elements* and found much, much to his liking. Al-Qiftī wrote many, many centuries after Euclid lived that, "Nay, there was no one even of later date who did not walk in his footsteps." Al-Qiftī is saying what I think is demonstratively true, we are all the disciples of Euclid.

The Life and Works of Archimedes
Lecture 6

> "These properties were all along naturally inherent in the figures referred to … but remained unknown to those who were before my time engaged in the study of geometry." –Archimedes.

A rchimedes was born in 287 B.C. in Syracuse, a city on the island of Sicily. He apparently lived most of his life in Syracuse, although there is some evidence that he might have studied at Alexandria. Much of our knowledge of Archimedes comes from Plutarch's *Life of Marcellus*, in which the mathematician figures as a supporting character. According to Plutarch, Archimedes was so engulfed in his work that he often forgot to eat or bathe.

These properties were always there, but it took Archimedes to see them.

Archimedes is known to some through the story of King Hieron and his crown. Hieron had given his goldsmith some gold to make a crown, but the king was suspicious that the goldsmith might have substituted a lesser alloy for the gold and asked Archimedes to help determine whether this was the case. According to the story, while bathing, Archimedes noticed that the water sloshed out of the tub when he lowered himself into it; he then realized that by lowering the crown into water, he could calculate its density and compare it to the density of gold to see if a substitution had been made. When he hit on this discovery, Archimedes is said to have leapt out of the bath and walked home naked, shouting "Eureka!"

Among the works that have come down to us from Archimedes are *On Floating Bodies*, *On Spirals*, and *Quadrature of the Parabola*, a book concerned with finding the area under a curve. His work entitled *Measurement of a Circle* has only three propositions, two of which are classics. In the first of these, he asserts that the area of any circle is equal to a right-angled triangle in which one of the sides of the right angle is equal to the radius and the other is equal to the circumference of the circle. We'll return to this proposition in the next lecture. In proposition 3 of this work, Archimedes states that the ratio of the circumference of any circle to its diameter is less than 3 1/7 but greater than 3 10/71. Recall that the ratio of the circumference of a circle (the distance around) to the diameter (the distance across) is a constant, π. In his calculations, Archimedes determined π to two decimal places, 3.14, although he expressed it in fractions rather than decimals. He accomplished this proof by approximating the circle's circumference by the perimeters of inscribed and circumscribed regular 96-gons.

Archimedes's masterpiece is *On the Sphere and the Cylinder*, in which he found the volume and surface area of a sphere. According to Archimedes, "If a cylinder is circumscribed about a sphere, the cylinder is half as large again as the sphere in volume and half as large again as the sphere in surface area." We can generate formulas for these amazing results using modern-day notation, and their accuracy can be confirmed through calculus.

Archimedes died during the siege of Syracuse by the Romans in 212 B.C. He had been given the job of defending the walled city against land and sea attacks ordered by the general Marcellus. Archimedes devised devilish weapons of war for Syracuse, against which the Romans were initially defenseless. When the Romans finally broke through the walls, however, Archimedes was found and commanded to appear before Marcellus. Because he was working on a proof at the time of his capture, Archimedes declined to follow his captor and was immediately slain. His memorial, according to Cicero, was a small column topped by a sphere contained in a cylinder. ■

Suggested Reading

Boyer, *The Concepts of the Calculus*.

Edwards, *The Historical Development of the Calculus*.

Heath, *A History of Greek Mathematics*.

Questions to Consider

1. A letter has survived from Archimedes to Eratosthenes, who was the librarian at Alexandria. Look up Eratosthenes, find out about his "sieve" for finding prime numbers, and read how he determined the size of the

Earth using only a few simple astronomical observations and Euclid's proposition I.29.

2. In proposition 34 of *On the Sphere and the Cylinder*, Archimedes stated, "Any sphere is equal [in volume] to four times the cone which has its base equal to the greatest circle in the sphere and its height equal to the radius of the sphere." Show that this yields the correct formula for spherical volume. Archimedes (of course) was right again!

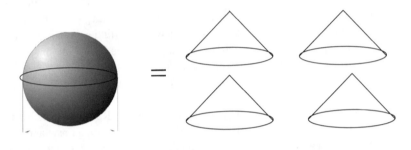

The Life and Works of Archimedes
Lecture 6—Transcript

In this lecture and the next we'll meet Archimedes who is universally acknowledged as the foremost of the Greek mathematicians. If I were building a mathematical Mount Rushmore, he'd be there alongside Newton, Euler, and Gauss. We'll meet them in later lectures. Now we want to meet Archimedes.

In this lecture, I want to talk about his life, his works, his personality, and his tragic, even iconic death. In the next lecture, we'll actually look at one of his great theorems where he determines the area of a circle. It's a piece of reasoning that's absolutely beautiful. So, we'll see Archimedes in action in the next lecture; but for now, it's time to look at his life.

Archimedes was born 287 BC in Syracuse, a city on the island of Sicily. It's still there in the shadow of Mount Etna. He apparently lived most all of his life in Syracuse, although there is some evidence he might have studied at Alexandria. Remember Alexandria, the great library was there, the School of Mathematics that Euclid had founded. We think he might have had some contact with that because there is correspondence that remains in which he writes to the mathematicians at Alexandria, and they seem to know him. So perhaps he spent some time studying there. But the fact is that most of his long life was spent in Syracuse.

We know a lot about him as a person due to *Plutarch's Lives*. Remember, Plutarch wrote this massive treatise about the lives of famous Greeks and Romans. Archimedes was not the subject of any one of these lives, but he figures as a supporting character in Plutarch's *Life of Marcellus*. A lot of the quotations I present in this lecture will be coming from Plutarch. So Plutarch tells us something about Archimedes as a person. Let me quote a passage. According to Plutarch, Archimedes would "… forget his food and neglect his person, to that degree that when he was occasionally carried by absolute violence to bathe or have his body anointed, he used to trace geometrical figures in the ashes of the fire, and diagrams in the oil on his body, being in a state of entire preoccupation… with his love and delight in science." So Plutarch here is portraying an absent-minded mathematician. He forgot to eat. He didn't bathe very much, and he was always being distracted by geometrical diagrams. Archimedes maybe wasn't the first absent-minded mathematician, but I can guarantee you he wasn't the last.

A famous story involving Archimedes has to do with the crown of King Hieron. Hieron was the king of Syracuse, the monarch in Archimedes' home town. The way the story goes, he had given his goldsmith some gold to make a crown, a royal crown. The goldsmith had returned the crown, but the king was suspicious that the goldsmith might have substituted a lesser alloy for gold. The challenge to Archimedes was to figure out if this crown was pure gold. Archimedes couldn't melt the crown down or anything, he had to use it intact and figure out how to determine its composition. The story goes Archimedes one day was in the bath, one of those rare occasions apparently when he did this, and as he lowered himself into the water, the water sloshed out of the bath; he realized that in this fashion, by lowering the crown into the water, he could eventually calculate the density of the crown, compare it to the density of gold, and see if in fact, there had been a substitution made. At this point, Archimedes, according to Vitruvius a later writer, "... moved with delight, he leapt out of the pool, and going home naked, cried aloud that he had found exactly what he was seeking. For as he ran he shouted in Greek: *eureka, eureka*!" So says Vitruvius. Here is an artist's rendition of that moment.

Two things are notable here. One is, even though a lot of us have great moments where we discover things, we usually don't forget to put on our toga. Archimedes was so smitten with this that he was running around without his clothes on. But secondly, this was the first eureka moment, a term that has entered the language as a moment of great discovery and excitement. So there we have it, something from Archimedes.

According to tradition, for his diagrams and calculations he would work on a sand tray that he carried around with him, a tray with sand in it. He would smooth it out, and then using a stick, draw his geometrical diagrams. It was sort of the laptop computer of its time, I guess. This is where he would do his scratch work, not having scratch paper as we do. This is very difficult. Imagine trying to do your mathematics on a sand tray. There are all kinds of things that can happen. The wind could blow away your proof or something.

It's amazing he did all that he did with his sand tray. Here is a picture of Archimedes working in the sand.

He contributed a whole lot to mechanics, to our understanding of mechanics. He studied the basic machines like pulleys and levers and screws. Just as an example is his famous statement, "Give me a place to stand and I will move the earth." What he's getting at there is he understood the principle of the lever, the fulcrum, if you had a long enough arm and you pushed down, you could lift a very great weight, even the earth, should you have a long enough arm to do that. His studies of the screw, the machine the screw, led to the Archimedean water screw for pumping fluids. This is a cylinder, an enclosed cylinder, with a spiral inside. As you turn the crank the spiral turns and the water rises from a lower to a higher point. This is still used. This is called the Archimedean water screw, a very nice application.

We have an awful lot of surviving works from Archimedes. We're actually quite fortunate in that regard, that so many of early mathematicians, their works are lost or contaminated. We have quite a number of Archimedes' works that are still around. Let me mention a few of these. One is a work called "*On Floating Bodies.*" This seems to be his work on hydrostatics, probably inspired by the story of the crown. He did a work called "*On Spirals.*" He was very much intrigued with spirals, and there is a spiral and spinning line that's still around called the Spiral of Archimedes. He wrote a book called, "*Quadrature of the Parabola.*" This name requires a little explanation. A parabola, as you know, is a curve. We think of it as being the graph of quadratic equation. Y equals X squared will graph to a parabola. The Greeks didn't have equations. They thought of the parabola as a conic section. That is, if you took a cone and sliced through it with a plane exactly parallel to one of the sides, it would slice out one of these parabolas. So that is what a parabola was to them, a conic section. A quadrature means "finding area." What he was doing here in this book was finding the area under a parabola, the area under such a curve. That's not a trivial question at all. Nowadays we do areas by integral calculus. If you look in this book, *Quadrature of the Parabola*, you can see Archimedes anticipating integral calculus. It's not quite there yet. He doesn't have the notation. He doesn't have the symbolism, but boy is he close. He is somebody who was ahead of his time not just by decades or centuries, but by millenia. So, that is great work indeed.

But then I want to mention two in particular, two of his works that you know, are sort of the top of the heap. One of them is called "*Measurement of a Circle*." It's a very short work. It only has three propositions in it, but two of them are real classics. The first is this: Proposition 1 in Measurement of a Circle he asserts that the area of any circle is equal to a right-angled triangle in which one of the sides about the right angle is equal to the radius, and the other to the circumference of the circle. This might sound rather confusing and complicated. This is actually his determination of the area of a circle. This will be the proposition we prove in the next lecture, so I'll say no more about it for now, but we'll be back. Proposition 3, the last proposition in Measurement of a Circle says that the ratio of the circumference of any circle to its diameter is less than 3 1/7, but greater than 3 10/71. Now you recall that the ratio of the circumference of a circle, the distance around to the diameter, the distance across, is a constant. Any two circles have that same ratio, whether it's a big circle with a big diameter or a small circle with a small one, the ratio is the same. That ratio of circumference to diameter we now denote by pi. That's what pi is. The Greeks didn't use that, even though that's a Greek letter. They didn't use that symbol, but they would always talk about the ratio of circumference to diameter. So what this is saying is, C/D, circumference to diameter is pi, and if you just cross multiply that, C is equal to pi D. There is that famous formula from high school. Whenever Archimedes is talking here about the ratio of circumference to the diameter falling between 3 1/7 and 3 10/71, what we can see that he's doing is saying that pi, this constant, lands between 3 10/71 below 3 1/7 above. If you get out a calculator and figure this out as a decimal, what this means is, he has nailed down pi to two decimal places, 3.14. He didn't express it as 3.14. The Greeks didn't have the decimal system. They had fractions. So the fractional expression that he gave would be good enough. But we can see this as sort of two-place calculation of pi. It's a major advance in geometry.

How did he do this? He did it by approximating the circle's circumference by the perimeters of inscribed and circumscribed regular 96-gons. Here is what he did. He took a circle. He wants that circumference, and he put inside it a regular hexagon, a regular 6-sided figure. He could tell the perimeter of the hexagon, which isn't the circumference of the circle, it's a little too small. But then he used a mathematical technique that allowed him, if he knew the perimeter of the hexagon, to double the number of sides to a 12-sided-gon, a dodecagon, and figure out that perimeter. Well, a 12-sided figure inside the circle is a little closer to the circle. Then he doubled from 12 to 24, doubled from 24 to 48, from 48 to 96. By the time you have a regular 96-sided polygon inside the circle, it's pretty close to the circle, and its perimeter is very close to the circle's circumference, but it's a little on the small side. He did the same on the other side. He circumscribed a hexagon, then a 12-gon, then a 24, 48, and 96-gon, and now sort of zooming in from the outside on the circle's circumference using that to get his upper estimate. By doing this, he has managed to come out with a very accurate estimate of pi, so long ago. This technique of doing this by putting circles in circumscribed and inscribed polygons to approximate the circle is called the classical method for approximating pi, due to Archimedes.

The other great work I want to mention is called "*On the Sphere and the Cylinder*." This was really his masterpiece. There he found the volume and the surface area of a sphere. This is a very sophisticated problem. The volume of a sphere, that's hard. The surface area, that's hard. Nowadays we do these with integral calculus. Archimedes didn't have that technique and yet he found these results. The big statement and his great result there, his great theorem is this, and I'm going to read this to you. He says, "If a cylinder is circumscribed about a sphere, the cylinder is half as large again as the sphere in volume and half as large again as the sphere in surface area." So again we get one of these very wordy kind of descriptions. At this point you're saying, why didn't he just give us a formula for this, for the surface and the volume of the sphere? Well, they didn't have algebra. As I've said many times, the Greeks were limited with their tools, at least it would seem to us, and so if you wanted to scribe a result, you have to put it in word; and this was his description of his great result.

When I see one of these, I have to ask myself, is he right? This doesn't look like what I know to be volume and the surface of a sphere. Is it right? When you ask this question to Archimedes, the answer is always yes. He is always right. It just doesn't maybe sound like it. It takes a little bit of work to see what he's getting at. So let me show you what this really is saying, this result.

First of all, he tells us he's taking a sphere and circumscribing a cylinder about. Here is the picture you usually see. There is the sphere, and when you circumscribe a cylinder, that means that you're going to take a cylinder whose radius is exactly the radius of the sphere, so on the vertical sides of the cylinder it just grazes the sphere; meanwhile its height, the cylinder's height, is the sphere's diameter, so that that circle on the top of the cylinder touches the top of the sphere, the circle at the bottom of the cylinder touches the bottom. That's what it means to circumscribe a cylinder about a sphere.

Well, now what Archimedes is saying is that the cylinder is half again the sphere in volume and in surface. What an amazing fact. Let's check these out individually. The volume of the cylinder—let's do the volumes first—the cylindrical volume, now you remember how to find the volume of a cylinder. You take the area of the base, which is a circle, times the height. So we usually write that V_c, the volume of the cylinder is πr^2, the area of the base times h, the height, so that's the cylindrical volume. But this isn't any old cylinder. This is a cylinder whose height is the diameter of the sphere as it goes from top to bottom. In other words, it's twice the radius. So, if I want to modify this formula, I'm going to say that the volume of the cylinder in this case is πr^2 times not h, but $2r$. If you multiply that out, you get $2\pi r^2$. So, the cylinder in question here has volume $2\pi r^2$.

Now, what does Archimedes say about this? He says this is half again the volume of a sphere in his sort of a verbal description of his result. What does that mean? Here is what he is saying. The volume of the cylinder is the volume of the sphere plus half the volume of the sphere again. It is one and a half spheres. You can see it's bigger. He says it's exactly one and a half of these. If I know this, add together V_s, plus half of V_s, I get three halves of V_s. Then if you cross multiply this, I'm trying to figure out what is the volume of the sphere according to Archimedes, you cross multiply and V_s will be 2/3 of V_c, is that three halves flips over. So, what he's saying is the volume of a sphere is 2/3 the volume of the cylinder, but up above we saw the volume of the cylinder was $2\pi r^3$. So what he's telling us in our notation is that the volume of a sphere is 2/3 of $2\pi r^3$, which is 4/3 πr^3. Is that right? You bet. Do we do it as Archimedes did? No, we do it with calculus. It's a nice calculus problem, but he's spot on there.

How about the surface? This seems to me even harder. How in the world do you figure out the surface area of a sphere? Well, he says the surface of the cylinder is half again the surface of the sphere. So, let's see what that's going to tell me. What's the surface area of a cylinder? Well, if you're going to paint the cylinder, what have you got to do? You've got to paint the circle on top, the circle on the bottom, and then the lateral surface of the cylinder, the distance around. So the surface area of the cylinder is the lateral surface, plus the top and bottom circle. You might remember the formula for the lateral surface of a cylinder. What you do is you imagine sort of snipping the cylinder up the side and opening it out. It has height h. It has width, the circumference of the original circle, $2r$. So the lateral surface of a cylinder is $2\pi rh$. Meanwhile, there is a πr^2 circle on the top, a πr^2 on the bottom, so this is the sum that will give me the total surface area of the cylinder, $2\pi rh + \pi r^2 + \pi r^2$. But again, this isn't just any old h. The h is $2r$, so my cylindrical surface as S_c is $2\pi r \times 2r$. Add those $2\pi r^2$ together, you get a $2\pi r^2$. Start collecting terms. You end up with $4\pi r^2$ for the first spot, plus $2\pi r^2$, altogether, $6\pi r^2$. So that's the surface of the cylinder. That's how much paint you'd need to paint the cylinder. Remember what Archimedes said is, that this is half again the surface of the sphere. That's going to generate a formula for the surface of the sphere in this fashion. You say the surface of the cylinder is the surface of the sphere plus half the surface of the sphere. As we saw in the previous example, that's going to end up being three halves of the surface of the sphere. Cross multiply, flip this over, and you get that the surface of the sphere is 2/3 the surface of the cylinder. But, we just said that the surface of the cylinder is $6\pi r^2$, so the surface of the sphere is 2/3 of that, which is $4\pi r^2$. Is that right? You bet it is. How do we do it? We do it with calculus.

So Archimedes is right on the money. It's really quite an impressive result. In fact, he stated it in a slightly different fashion, which I like, he said, "The surface of any sphere is four × its greatest circle." That's almost poetry there. "The surface of any sphere is four × its greatest circle." What does he mean by that? If you wanted the surface, say, of a basketball, there is a sphere. What's the surface area? How much paint would you need to paint that? He says, slice through this, so as to get the greatest circle, which if this were the earth, you'd slice it right through at the equator. I slice through there, and then imaging pulling that greatest circle out and putting it flat, and taking a second, and a third, and a forth. What he has proved is, the surface of the sphere, this curved surface in three dimensions, is exactly the sum of those four flat circles. That's quite amazing. Not kind of equal to four of them, but exactly equal to four of them. The surface of a sphere is $4\pi r^2$. That's Archimedes. Now he says at some point that, "... these properties were all along naturally inherent in the figures referred to," he didn't create these, they were always part of the spheres, he said, "but remained unknown to those who were before my time engaged in the study of geometry." These properties were always there, but it took Archimedes to see them. So, it was an amazing mathematical legacy.

In the time that remains, let me tell you about his death, which is a famous story that mathematicians revere. This occurs during the siege of Syracuse in the year 212 BC. Recall at this time Rome was expanding its borders. Rome was growing. They had attacked Carthage. They had attacked Greece. Now they were looking at the island of Sicily, and in particular, the city of Syracuse. The Roman general assigned to capture Syracuse was named Marcellus, and of course it was he that Plutarch was writing about whenever Archimedes gets into the story.

Marcellus sends troops by land and ships by sea to capture Syracuse, which was right on the coast, figuring that one way or the other he was going to get over the walls, get inside, and capture the city for the glory of Rome. However, inside Syracuse was old Archimedes, and he was given the job of trying to defend his walled city against both the land and sea attacks that were coming from the Romans. So Archimedes began creating these devilish weapons of war to keep the Romans out. Plutarch said all of the Roman efforts were, as it were it would seem, but trifles for Archimedes and his machines, his armament.

Here come the Romans on the land side, and what happens? Well, according to Plutarch, is "...when Archimedes began to ply his engines, he at once shot against the land forces all sorts of missile weapons, and immense masses of stone that came down with incredible noise and violence, against which no man could stand." So the army is marching to the walls, and here come stones and arrows and all sorts of things from Archimedes' wonderful machines, and the Romans ran away. Now Marcellus needs another strategy. Let's send the Navy in from the other side. That didn't work either because in the meantime huge poles were thrust out from the walls over the ships, sunk some by the great weights they let down from on high. Apparently the boats would be under the wall and this big arm would come out of a big stone on it and just sink the ship; or Plutarch says that other times this claw would come out over the wall that Archimedes had somehow invented, and it would go down and grab the ship and then pick it up and shake all the sailors off into the water, which was terrible. Now the Navy had been thwarted as well. The Romans were not getting anywhere against Archimedes and his weapons.

According to Plutarch, "…when such terror had seized upon the Romans… if they did but see a little rope or a piece of wood from the wall … they turned their backs and fled." They were just terrified of what was next. What else was coming over this wall? Plutarch says, "…the Romans, seeing that indefinite mischief overwhelmed them from no visible means, began to think they were fighting with the gods." So Archimedes was at least initially successful in keeping the Romans out. Who knows how long this would have gone, but as the story goes, there was a festival day at one point, when the people in Syracuse, the guards, started drinking a little too much, let down their vigilance, and the Romans broke in. They got into the walled city. Now they're running around through Syracuse. Archimedes is in there, but he's not paying attention. He's working on a math problem. Plutarch says that Archimedes "…having fixed his mind upon working out some problem … never noticed the incursions of the Romans nor that the city was taken…" So he is busily working on his math problem. Here come the Romans to attack. A soldier unexpectedly coming up to him commanded Archimedes to follow to Marcellus. The soldier finds Archimedes and says, "You're coming with me back to the general."

Marcellus would have loved to have had Archimedes on his side to invent all these weapons of war for the Romans. Here is a mosaic from classical times that shows this scene. You see the centurion there. You see Archimedes. He is actually working on his sand tray, as he did. The centurion is saying, "Come with me back to Marcellus."

According to the story, Archimedes declined to do this before he had worked out his problem to a demonstration. "No, no," he says, "I'm not going with you. I've got to finish my proof." Wrong answer because then the soldier enraged, drew his sword and ran him through. Archimedes was slain because he wouldn't stop doing his math. Here is an artist's rendition of this.

Ever since, mathematicians have talked about this as the great mathematical death; what a way to go. Here you are working on your problem; you won't give it up until you get your proof and, okay, that cost you your life. This story has intrigued people for generations. There is a little more to it. Plutarch said that Archimedes "...is said to have requested his friends and relations that, when he was dead, they would place over his tomb a sphere contained in a cylinder." What he asked was as his memorial, the very diagram of the sphere within the cylinder that had led him to this great result, his real masterpiece, and pretty clearly he saw it that way as well, that was to be his memorial. Then the story goes, a few centuries later when Cicero, the Roman orator, is down in Sicily visiting, and he us tells that amid a jumble of brambles and bushes, "I found," said Cicero, "a small column that emerged a little from the bushes. It was surmounted by a sphere in a cylinder." So according to Cicero, as he was rooting around in the brambles near Syracuse, he found the tomb of Archimedes with that very symbol atop. Then what Cicero said is that cleaned out the brambles. He tidied up the area. You can see this as an homage to the great Archimedes, and perhaps, even an apology from the Roman Cicero to the Greek Archimedes, because after all, by then, Greek civilization had been brought under the heel of the Romans and would never again be quite what it once had been.

Let me finish this with a few words from Plutarch on Archimedes. Plutarch says, "It is not possible to find in all geometry more difficult and intricate questions, or more simple and lucid explanations ..." "No amount of investigation of yours would succeed

in attaining the proof, and yet, once seen, you immediately believe you would have discovered it; by so smooth and so rapid a path he leads you to the conclusion desired." According to Plutarch, Archimedes' words were next to impossible to imagine you doing it. You'd never have gotten it until you see how he does it and then you say, "Why yeah, I'd have figure that out." Well no, you wouldn't have figured it out, because you weren't Archimedes. But Archimedes left behind this wonderful legacy of great results that seemed a lot superhuman. That's why he's on my Mount Rushmore.

In the next lecture, I'm going to show you one of these results, and you'll see whether you believe that he smoothly and rapidly led you on a path to the conclusion desired.

Archimedes's Determination of Circular Area
Lecture 7

Recall from the last lecture the first proposition of Archimedes in *Measurement of a Circle*: "The area of any circle is equal to a right-angled triangle in which one of the sides about the right angle is equal to the radius and the other to the circumference of the circle." What he's really saying is that a circular area is the same as that of a particular triangle.

The Greeks often used the strategy of comparing a complicated figure to a simpler one. In this case, the circle is the more complicated figure and the triangle is the simpler. To find the area of a triangle, we return to Euclid's *Elements*. Proposition I.41 states, ""If a parallelogram have the same base with a triangle and be in the same parallels, the parallelogram is double of the triangle." Unpacking this statement, we find the formula $1/2(bh)$ for the area of a triangle. In Archimedes's triangle, the base is c, the circumference of the circle, and the height is r, the radius; thus, the formula for the area is $1/2(cr)$.

To proceed further, Archimedes had to establish two preliminary results, often called lemmas in mathematics. Think of a lemma as something we prove because we will need it later. First, Archimedes wanted to determine the area of a regular polygon based on a measurement called the apothem (the perpendicular distance from the center of the polygon to any of the sides) and the perimeter. The formula for this is 1/2 the product of the perimeter and the apothem. The second lemma needed can be stated as follows: Given a circle, we can inscribe within it a square, then double the number of sides and get a regular octagon, then double the number of sides and get a regular 16-gon, and continue until the difference between the circle's area and that of the inscribed polygon is as small as we wish. This approach is sometimes called the method of exhaustion. In modern language, we would say that the limit of the areas of these polygons approaches the area of the circle. Note, however, that there is no polygon whose area is exactly the area of the circle from the inside because polygons have straight sides, while the circle is round.

With these lemmas, Archimedes was able to prove the original proposition using a method we might call double *reductio ad absurdum* or a double contradiction. He first assumes that the area of the circle is greater than the area of the triangle. Using the two lemmas, he finds, first, that the area of a polygon inscribed within the circle is greater than that of the triangle; then he finds that this same area is less than that of the triangle. Having reached a contradiction, he eliminates this first assumption; he then uses a similar process to eliminate the assumption that the area of the circle is greater than that of the triangle. Finally, Archimedes states that if the area of the circle is neither greater nor less than the area of the triangle, the two are equal. We can trace Archimedes's proof to find our familiar formula for the area of a circle: πr^2. We can also take a more direct approach to determining the area of a circle using integral calculus, which simplifies the process into a series of steps and allows us all to become mathematicians like Archimedes. ■

Suggested Reading

Boyer, *The Concepts of the Calculus.*

Edwards, *The Historical Development of the Calculus.*

Heath, *A History of Greek Mathematics.*

Questions to Consider

1. We saw that the area of a regular polygon is half the product of its perimeter and its apothem. As a formula, this becomes: area $= \frac{1}{2}Ph$.

Suppose our regular polygon is a *square*. Show that our formula yields the correct area in this (simple) case.

2. As we saw, Archimedes established his formula for circular area by the logical strategy of double *reductio ad absurdum* (double contradiction). Conjure up a situation—it need not have to do with mathematics—in which a result could be established by **triple** *reductio ad absurdum*.

Archimedes's Determination Of Circular Area
Lecture 7—Transcript

In the previous lecture, we met Archimedes, a great thinker. In this lecture, I want to look at a great theorem that he contributed. We want to look over his shoulder as he determines the area of a circle and one of the most amazing proofs I know.

Before we get to this, let me say that this lecture more than any of its predecessors is highly mathematical. The whole lecture, in fact, is going to be devoted essentially to this one proof. It's going to be a bit challenging, but here's my pledge. I will try to break it down into easy-to-follow steps. The hope is that by the end you will understand and you will be able to appreciate this genius, who's working before us, over whose shoulder we are looking. That's the object; let's begin.

The issue, as I said in the previous lecture, is addressed by Archimedes in his work *Measurement of a Circle* from about 225 B.C. It's the first proposition that I want to look at and I shall read it to you one more time. Here's what it says: "The area of any circle is equal to a right-angled triangle in which one of the sides about the right angle is equal to the radius, and the other to the circumference, of the circle." It's a very strange statement here. What's going on here with all these words? What he's really saying is that a circle, a circular area, is the same as that of a particular triangle. Imagine we have a circle—here it is, radius R—and he says let's build a right triangle; it's going to have a right angle in it. There it is. One of the sides about the right angle is R, the radius, and the other is the circumference of the circle. If you imagined snipping the circle and straightening it out that's going to be the other side. We have a circle, we have a triangle, and he says they're equal in area.

This strategy is something that the Greeks generally, and Archimedes particularly, would employ. You want to compare a complicated figure to a simpler one. We saw this in the last lecture and remember when he was looking at the surface area of a sphere, complicated, he says that's four times the greatest circle within. Circles are simpler than spheres, so the complicated spherical surface is cast in terms of the simpler circles. Or, he talked about the volume of the sphere, complicated, and compared it to the volume of a cylinder simpler, simple. Also here, the circle this time is the complicated figure; he says it's going to equal this triangle and triangles are much easier than circles. This was the way the Greeks would do this; they would compare the complicated to the simple.

I say the triangle is simple, but maybe we better stop and make sure we believe that. What's the area of a triangle? If I'm going to say the area of the circle equals the area of the triangle, I better be sure I'm on top of triangles. That's not hard. This shows up—guess where—in Euclid's *Elements*. You thought you were done with Euclid, but he's back again. He always is. This is Proposition 41 of Book 1 where Euclid proves this proposition. He says: "If a parallelogram have the same base with a triangle and be in the same parallels, the parallelogram is double of the triangle." Again, it's not a formula, but it's a sort of wordy statement. Let's see what he's getting at and see if we can unpack this and see what's up.

There's a triangle, let's say its height is h, its base is b, and what Euclid is saying is put it within a parallelogram so that the triangle and the parallelogram have the same parallels and the same height. In this case, I'll just put it inside a rectangle. There's a nice simple parallelogram. The triangle and the rectangle fall within the same horizontal parallels and they have the same base. According to Euclid, Book 1, Proposition 41, the area of that rectangle is double the triangle, two triangles. The area of the rectangle, that's easy, base times height, so that's two triangles. Divide everything by two and what this proposition is saying is the area of a triangle is $1/2bh$. There's a famous formula we all know, triangular area is half the base times the height. Where'd it come from? If I were asked that, I'd say it comes from Euclid's *Elements*. But, as you can see it wasn't exactly expressed this way; again, the limitations of the Greeks not having algebra they had to give you a verbal statement, but you can rather easily see that's what Euclid really was getting at.

Areas of triangles we know. Back to that one Archimedes was looking at, remember he said the circle equals that triangle, one side is the radius, one side is the circumference, so what is the area of the triangle? It's $1/2bh$. For that particular triangle he's talking about the base is C, the circumference, and the height is R. The triangle in question has area $1/2CR$. That'll be important to remember as we walk through this proof.

To proceed further, Archimedes needed a pair of preliminary results, two things he has to establish. They're not the proof, but there're going to be tools needed in the proof. Sometimes in mathematics, these are called lemmas, L-E-M-M-A. A lemma is something you prove because you're going to need it later. We have two lemmas if you will for this argument. Let's take a look at these.

The first one involves regular polygons. We've encountered them before; remember polygons are regular. If all their sides are the same length, all their angles are the same. They're these beautiful symmetrical polygons. What Archimedes is trying to do in this first lemma is figure out the area of this thing based on other measurements. In particular he wants to introduce something called the apothem. The apothem might be a term you're not familiar with, but it is the perpendicular distance from the center of the polygon to any of the sides. What you'd have to do first is find the center of this regular polygon; you could just draw two diagonals and where they cross will be the center. Then you go out perpendicularly to any side, because of the regularity you have the symmetry that you can go in any direction you want, and that length will be the apothem. In my picture I've drawn my perpendicular from the center upward and that length is h. The apothem is going to figure in this.

The other thing he's going to need is the perimeter. That is much more familiar; the perimeter of a regular polygon is just the distance around. If you were taking a stroll around the polygon, that's how far you would travel. What he's seeking is an expression for the area of the regular polygon in terms of these two quantities, the apothem, the perpendicular distance out to any side, and the perimeter, the distance around. The Greeks knew this. I don't know if modern people are familiar with this, but let me show you what this formula is.

There is my regular polygon; I've drawn in the apothem h. The way to derive this formula is to draw it from that center, two lines to two adjacent vertices. Let me do that. There are two dotted lines there that form a little triangle, thus formed. The height of this triangle is h and the base of the triangle, the horizontal side, let me call that x. I've put in a little wedge, if you will, a little triangular wedge inside my regular polygon. The area of that triangle—we just saw that from Euclid—the area is $1/2bh$, it's going to be $1/2xh$. That's the area of that particular triangle.

Now what you do is you draw other lines emanating from the center to the other vertices. There's a second, a third, and a fourth; I have an octagon here so I'm going to have eight of these, but it could be any number. You've put in lots of little wedges and the bases of each of these triangles thus formed are the same, x, x, x, all the way around because it's a regular polygon—all their bases are equal. This picture is sort of the decomposition of the regular polygon into the various little triangles and now what you say is the area of the whole polygon, all the way around, is just the sum of the triangular wedges shooting out from the center. But, each of those is the same as this one we did up above, $1/2xh + 1/2xh + 1/2xh$ all the way around until you get to the end, how many ever there are. That's the area of my regular polygon.

Now what I'm going to do is factor out a $1/2$. Every single term in that expression has a $1/2$. I'm going to pull the $1/2$ out to the left and I'm going to factor out an h, every single expression has an apothem in it, an h in it, I'm going to factor that out to the right. When I do that, I'm going to be left with this expression $1/2$ the quantity $x + x + x + x$ finally times h; $1/2$ out one side, h out the other, and all those x's added up in between. If you add up all those x's, x, x, x, you're just going around the polygon and so the sum in the middle in the parenthesis is just the perimeter. There is the first result we're going to need. The area of a regular polygon, if I want to put it in words, is $1/2$ the product of the perimeter times the apothem. We'll see Archimedes using this in a minute.

But, there's another preliminary result he's going to need and it is a little more wordy. Let me state it this way: Given a circle you can inscribe within it a square and then double the number of sides and get a regular octagon, and then double the number of sides and get a regular 16-gon, and continue until the difference between the circle's area and that of the inscribed polygon is as small as we wish. You can put within a circle a square, 8-gon, 16-gon, as it sort of fills up the circle from within, you can make the difference between the circular area and the polygonal area as small as you like. You can do the same with circumscribed polygons. You can put a polygon around the circle, say a square around the circle, and then an 8-gon, an octagon, a 16-gon and sort of approach it from outside in and make the difference between them an area as small as you wish.

This idea is sometimes called the "method of exhaustion." You're exhausting the circle from within; you're exhausting it from without. This predates Archimedes. This was due to a Greek mathematician named Eudoxus from around 37 B.C. Eudoxus introduced the method of exhaustion, which is quite sophisticated, but Archimedes was the real person that exploited it so nicely. I want to show you a picture of what it is he's saying here. Maybe my verbal description isn't as good, but I can show you this and you can believe it. Let's suppose we have a circle. There we go. I'm going to inscribe within it a square. I have my blue circle in which I'm putting a square. Notice the difference between these two areas is that excess blue around the outside and there's kind of a lot of it right now—I'm not very accurate here; my square doesn't fill up the circle very well. Now what I'm going to do is take twice as many sides, go to an octagon. You can actually do this with a compass and straight edge. If you wonder how to do it, you would take the side of the square, you'd bisect it, and then put up a perpendicular. Where it hits the circle, you now draw two sides where you had previously one, so instead of four you get to eight. That process can be repeated to get to the 16 and the 32 and all that.

In my picture, there's my circle. Now I'm putting in an octagon and notice there's still an excess of blue over white. The circle's a little bigger, but the difference is much smaller. What Archimedes needed was the fact that we can make that blue ring on the outside as small as we want by just repeatedly doubling the number of sides by exhausting the circle from within. If I want the inscribed polygon to be within one-half a square inch of the circle, I can do it by taking enough sides. If I want it to be within one one-millionth of a square inch of the circle, I can do it by taking enough sides. I can make the polygon as close as I want to the circle. If you've seen calculus, this certainly sounds like the notion of limits. What we would say in modern language is the limit of the areas of these polygons is the area of the circle. It is approaching it as closely as we want. One thing we should note though, it never gets there. There's no polygon whose area is exactly the area of the circle from the inside because polygons always have straight sides even if there are millions of them whereas the circle is round, so it's not going to ever reach it, but the important thing is we get as close as we want. The picture I think shows that.

Let me show it from the other side. How about the circumscribed approach? There's my circle. I'm going to circumscribe about it a square, so there's the circle and I put the blue square around it. Obviously there's excess blue; there's much more area in the square than the circle. All that blue stuff is the excess. Now if I go from the square to the octagon, I will see the picture look like this. There's the circle, here comes the octagon, and it's much less excess blue. I've gotten closer to the circle's area. Do it again and go to a 16-gon, closer yet, 32-gon, closer yet. We have a limiting process going on from the outside. What Archimedes needs is that no matter how small a gap. you can make the polygon be an inscribed or circumscribed within that gap to the circle's area.

Those are the preliminaries. Here we go; now we're ready. The lemmas are out of the way. Let's see how he does this. The proposition, remember, was that the area of the circle equals the area of the triangle. The circle had radius R, the triangle was this very special one with one side R and the base C, the circumference. Here comes a surprise. How do you prove that the circle equals the triangle? He does it in a most strange fashion. He does it using what we'll call double reductio ad absurdum. Teductio ad absurdum, reduction to an absurdity, that's the sort of Latin term for contradiction. He's going to show that this circle equals this triangle by double contradiction. It's a real surprise for the readers the first time they see this and wonder why he's doing this, but just hang on because this is beautiful stuff.

What he does first is, in case 1, remember he's going to end up getting two contradictions here. First case, suppose the area of the circle is bigger than the area of the triangle. That's not what he wants to show. He wants to show the area of the circle equals the triangle, but let's suppose for the sake of argument that the circle is larger. Let me draw a picture like that. There's the circle—I've drawn it really large—and there's the triangle with sides r and

C. My picture makes it pretty clear that the circle is bigger. This is the kind of thing he's considering in this case.

Archimedes says let's inscribe within the circle, inscribe within the circle, a square and then an octagon and then a 16-gon and so on using his second lemma, until he gets the area of the circle bigger than the area of this inscribed polygon, which in turn is bigger than the triangle. I'm going to say a word about it, but let me show you the picture of how this could work. I've put an octagon in there, but it might take more than eight sides; it might take 800 sides or who knows what. In any event the area of the circle is bigger than the area of the polygon inscribed within it, which is bigger than the triangle. Can he do this? Is this legit? Yes, because remember what the second lemma said—if you give me the area of the circle, I can make the area of the inscribed polygon as close as I want, as close as you want. How close do I want it? I want it to be closer to the circle's area than that of the triangle. Remember the circle was bigger than the triangle, so there was some gap. The triangle was down here; the circle's up here. I can do my inscribed polygon so they get as close as I want; all I need is to make sure the polygon is closer to the circle's area than was the triangle as my picture indicates. He's exploited that second lemma.

Now what? Look at that circle with the inscribed polygon. Let's take a look at the apothem. You've got to remember what's that. It's the line from the center perpendicularly out to one of the sides of the polygon. Then there's the radius, which is the distance from the center out to one of the vertices. I've drawn in the radius here; there comes the apothem. You can see the two lines going out from the center—the apothem perpendicular to the side, the radius out to the corner, the radius of the circle in which this polygon is inscribed. Look at the picture. These aren't the same length. One is shorter and one is longer; clearly the apothem is shorter than the radius. Actually, if you look at that little triangle that appeared there, the radius is the hypotenuse of the right triangle and the apothem is a leg and the hypotenuse is longer than any leg. What I have observed is h is less than r; the apothem is shorter than the radius. Store that information.

How about we compare the perimeter of the polygon in there to the circumference of the circle. Which is shorter? If you had to walk around the polygon or walk around the circle, which would be the shorter journey? Since it's inscribed, since your polygon is inside the circle, the P, the perimeter of the polygon, is less than C, the circumference of the circle. From that, we make this conclusion. The area of the polygon, remember Archimedes's first preliminary, it's $1/2P$ times apothem, $1/2Ph$. There we're going to exploit that. But, wait a minute, h is less than R, we just said. P is less than C, so $1/2Ph$ is less than $1/2Cr$. At this point you're thinking where is this going? Is he adrift? Is there an end to this? Yes, it's very close actually. In fact, it's staring you in the face. There is a contradiction right in front of you. I don't know if you see it, but let me show you where this is. There's trouble here, logical trouble. Look at the line above. I have there that the area of the polygon is greater than the area of the triangle. Let me put a circle there. But what I just showed was the area of the polygon, which is less than $1/2Cr$, which was the area of the triangle is less than the area of the triangle.

Look at these two lines—the area of the polygon is more than the area of the triangle and the area of the polygon is less than the area of the triangle. That can't be. That is a logical contradiction, an impossibility. What we conclude is by jumping into Case 1 and deducing all of this we've reached a contradiction—Case 1 is impossible. Throw it out; it can't be. Remember what Case 1 said? The circle is more than the triangle. We've just eliminated that. That's one contradiction.

Case 2: Archimedes says maybe the circle is smaller than the triangle. He doesn't want this either. He's hoping to eliminate this option as well, and he does and here's how. If the circle is smaller than the triangle, let me draw the picture where the circle is little and the triangle is big, so I make it look like the circle is certainly smaller than the triangle. It's that same triangle, r and C being its dimensions. Now what Archimedes does is exploit the second lemma and says that we're going to circumscribe about the circle a square, then an octagon, then a regular 16-gon, until the area of the circle is less than the area of the polygon, is less than the area of the triangle. My picture would show this polygon surrounding the circle, whose area is certainly bigger than the circle's, but smaller than the triangle's. Here we go again with this lemma. Remember the circular area is here. We said the triangle's bigger than the circle. What I want to do is take a polygon surrounding the circle, circumscribed, until its area gets so close to the circle's that it's less than that level at the triangle. That's what we've got, so that is legitimate by the second lemma.

Back to the circle on the polygon, let me draw the radius of the circle r and compare it to the apothem as we did in Case 1. In this case, the radius goes from the center perpendicular to the side of the polygon. It's exactly the same as the apothem. In this case since we're circumscribed, the radius goes out to the circle, the apothem goes out to the polygon, but it's going out the same distance. In this case h is r; we don't get an inequality. But, we do when we compare the perimeter of the polygon to the circumference of the circle. Which is bigger? Again, if you had to take the journey around the polygon or around the circle, which is the further walk? Since we're circumscribed this time, the polygonal perimeter is more than the circular circumference, so P is bigger than C.

From that we deduce that the area of that polygon by Archimedes's first lemma is 1/2 perimeter times apothem. There that comes again. But, the perimeter we said is bigger than C, the apothem h is equal to r, so $1/2Ph$ is bigger than $1/2Cr$ and $1/2Cr$ is the area of the triangle as we saw earlier, $1/2Cr$. Again, a contradiction is staring you in the face. Can you see it? In the line above, we said the area of the polygon is less than the area of the triangle; in the line below, we said the area of the polygon is more than the area of the triangle. That can't be. They can't both hold, something's wrong; we've got a contradiction. We therefore can eliminate Case 2 as impossible.

We're almost there. Here's what Archimedes said. Case 1, he showed it was impossible that the area of the circle is more than the area of the triangle. That disappeared in Case 1—gone, can't be. Then he showed it was impossible that the area of the circle is less than the area of the triangle. That went down the tubes in Case 2. What's left? Archimedes puts it this way—since then the area of the circle is neither greater nor less than the triangle, it is equal to it. That's his result. That's his proof by double contradiction, which seems to me to be really quite amazing—a eureka moment. There we go. That's it; that's what he wanted to prove.

A modern reader looks at this and says this is strangely indirect. What a strange way to prove that two things are equal. To show that one can't be bigger than the other or smaller than the other, hence they must be equal. We sort of eliminate all the other possibilities. It seems peculiar. Here would be an example. Suppose somebody said let's prove $2 + 3 = 5$, which it does, right. The Archimedean approach, the double reductio ad absurdum approach, would say let's first assume $2 + 3$ is more than 5 and you do some reasoning and you get a contradiction, so that can't be. Then you assume in Case 2, $2 + 3$ is smaller than 5, and you do some reasoning and get a contradiction. That's out. If it can't be bigger than 5 or it can't be smaller than 5, you would be justified in concluding that $2 + 3$ is 5. What a strange route to that end. It's surprising, but as I've said before the history of mathematics holds interesting surprises.

Here are two other things. First of all, is he right? We're back to this question. The area of the circle equals the area of that triangle. We proved it, but what's this got to do with the formula that I know for the area of a circle? It doesn't look like it. I don't learn it this way. That's easy to fix. The area of the triangle we said is $1/2 Cr$, 1/2 base times height—but C wasn't just anything, remember? It was the circumference of the circle. C is the circle circumference. We've seen that C/D is Pi, so the circumference is Pi times the diameter, but the diameter is $2r$ so C is $2\pi r$, a famous formula. Stick that in there where the C is and you'll see that what he's telling us is the area of the circle is $1/2(2\pi r)r$, cancel the 2s, and the area of the circle is πr^2. That's Archimedes's result in our language. If you ever see that, area of a circle is πr^2, where did it come from? It came from right here, Archimedes's measurement of a circle.

Let me conclude with how a modern mathematician would do this. How would we determine the area of a circle today? In particular, how can we avoid this roundabout method of double reductio ad absurdum by sort of doing it by eliminating the other cases? Is there a direct attack and a direct assault on circular area? There is, but it's going to take integral calculus. Again we see Archimedes ahead of everybody, as he is sort of anticipating integral calculus.

If you don't know calculus, just hang on for a minute because I want to make a point here about this derivation, but if you'll do you'll recognize some of the symbols. Let me just show you how this works. I take a circle, I want to insert the axes; the origin is there. Let's say the circle is going to have radius r, so its r units out to one end of the radius, r units up. If you're doing this with calculus, you then need the equation of that circular arc, which you can determine in a pre-calculus course to be y equals the square root of r squared minus x squared. Then, the calculus argument would say, let's go after the area of that blue region. That's sort of the upper quadrant, a quarter of the circle. Let's figure that out. Calculus says that the shaded area is an integral. It's the integral from 0 to r of ydx. But, y we said is the square root of r squared minus x squared. The problem as a calculus problem gets converted to this thing—can you evaluate the integral from 0 to r of the square root of r squared minus x squared dx?

You can do this if you know calculus. You spend weeks and weeks and weeks doing things like this. Let me just quickly tell you what happens here. This comes out to be the anti-derivative is $1/2r^2 \arcsin x/r + 1/2x\sqrt{r^2}(x^2)$, and then the calculus course teaches you to evaluate this between 0 and r. If I put in r for x, I end up having to know the arcsin of 1, which is $\pi/2$, everything else cancels out, and so I get $1/2r^2(\pi/2)$ and the shaded area therefore is $1/4\pi r^2$. You finish up by saying that the shaded area is $\pi r^2/4$. Therefore, the whole area of the circle is four of those—$4\pi r^2/4$ is πr^2.

The details of this aren't so important, but what's important is that this is a direct attack on the problem. I didn't have to eliminate the other candidates; I went straight at it and got the area of the circle is πr^2. That's what calculus does. That's why calculus is such a great advance over what came before. It gives you this direct assault on problems like this and so many others. It's got another advantage. What I showed you there, this calculus derivation of the circular area, is just a bunch of steps; you just follow the rules from one to the other and you can do it. You don't have to be brilliant like Archimedes and figure out the roundabout way to do this. Calculus turns all of us into little Archimedes who can do these very difficult problems. Calculus sort of sold itself, it made a name for itself, by arguments like this that could turn a complicated problem into a simple one.

Archimedes was certainly great in spite of the fact that he was centuries before calculus. I'll leave you with a quotation from Voltaire celebrating the great Archimedes. Voltaire in the 18th century wrote these true words: "Archimedes had at least as much imagination as Homer." Archimedes was really great.

Heron's Formula for Triangular Area
Lecture 8

I challenged you to find the square root of 336. ... I'll give you nothing but a stack of paper and a pencil or maybe a stack of papyrus if we want to be in the spirit of Heron. How do you do something like this?

Our last Greek mathematician is Heron of Alexandria (sometimes called Hero of Alexandria), who is usually dated to around the year 75. Heron is credited with inventing the aeolipile, a proto-steam engine, and with devising a method for drilling a tunnel from both sides of a mountain at once. His great work was the *Metrica*, in which we find his method for approximating square roots and his formula for triangular area.

Suppose we want to find $\sqrt{336}$, with Heron's method, we begin with an estimate (x_1). We'll try 18; 18^2 is 324. For a second estimate (x_2), we take x_1 divided by 2 + the number whose square root we seek divided by $2 \times x_1$; in this case, the expression would read $18/2 + 336/2(18)$, which yields 55/3, or 18 1/3. We then repeat the process to get a third estimate, which yields 6049/330, or 18 109/330. If we square this result, we get very close to 336. The modern idea of limits shows us why this technique works. The general idea is as follows: Let x_n be the approximation at any stage of $\sqrt{A}$. The next approximation, $x_n + 1$, will be $x_n/2 + A/2x_n$. As n goes to infinity, we see that x_n approaches $\sqrt{A}$.

An even greater result in the *Metrica* is the formula for finding triangular area without knowing the altitude of the triangle. For a triangle with three sides, a, b, and c, we first find the semiperimeter, s, which is half of the perimeter. Heron's formula for the area of the triangle then is $\sqrt{s(s-a)(s-b)(s-c)}$. This formula seems implausible because it is completely unrelated to the formula we usually use, $1/2(bh)$, but if we substitute values for a, b, and c, we can see that it works.

Suppose we had a four-sided plot of land measuring $10 \times 17 \times 25 \times 24$ yards. There is no unique answer for the area of this quadrilateral. Just knowing the side measurements of a four-sided figure does not determine the area, but we can find the area of a three-sided figure using the side measurements. Thus, if we draw a diagonal in the quadrilateral, we can find its area by finding the area of the two resulting triangles. In this case, say the diagonal measures 26 yards. Using Heron's formula, we determine the area of one triangle to be 120 and the second triangle to be 204; thus, the area of the four-sided plot of land is 324 square yards. In fact, we can break any polygon into triangles, apply Heron's formula repeatedly, and sum the results to find the total area.

Heron's proof of this formula is quite complicated. As we will see in a later lecture, Isaac Newton offered a much simpler proof.

Heron's proof of this formula is quite complicated. As we will see in a later lecture, Isaac Newton offered a much simpler proof. Interestingly, Heron's formula implies the Pythagorean theorem as a consequence, which we can see by finding the area of two congruent triangles using the usual formula and Heron's formula.

The Greeks were impressive in their ability to delve deeply into mathematics with limited tools and without many of the modern mathematical results that we take for granted. With the work of Thales, Pythagoras, Euclid, Archimedes, and Heron, the Greek mathematical legacy is unsurpassed. ∎

Suggested Reading

Heath, *A History of Greek Mathematics*.

Questions to Consider

1. Starting with a first approximation of $x_1 = 10$, use two applications of Heron's technique to approximate $\sqrt{105}$. How accurate is the estimate?

2. Use Heron's formula to find the area of a triangle whose three sides are of length 25, 52, and 63.

Heron's Formula for Triangular Area
Lecture 8—Transcript

This will be my last lecture on a Greek mathematician and for this one I've chosen Heron of Alexandria. I want to tell you about his work and particularly about his amazing formula for triangular area.

Heron is often called Heron of Alexandria. That's where he worked, that same place which had been so prominent in the history of mathematics where Euclid had been Heron eventually was. Strangely we know very little about him. Even though he was much later than Archimedes, we know much less about Heron. Maybe that's because he didn't have the privilege of being killed by a Roman soldier, but in any event, his life is kind of obscure. We don't even know his dates very well to within decades, but he's often placed at around the year 75, so let's say we'll go with that for Heron.

I should say that you might see him in the literature called Hero, so sometimes you'll see references to the mathematician Hero of Alexandria and Hero's Formula. In case you do I want to explain what that's about. A friend of mine of Greek extraction explained this. The Greeks had a famous philosopher named Platon, but Platon when his name got Latinized became Plato—the N disappeared. The very same phenomenon occurred with Heron. Heron was the Greek mathematician; when his name got Latinized, the N dropped off and he became Hero. To be consistent I should either talk about Platon or Heron or about Plato and Hero, but I'm going to talk about Plato and Heron, so go with that. I will say this, however, I'm glad that this phenomenon ended, that they stopped dropping the N, or otherwise Newton would've become Newto and that would definitely not have worked.

In any case, what were Heron's achievements? He had many. They were often of an applied nature rather than a theoretical one. For instance, he is credited with inventing the aeolipile. What's that? It's a proto-steam engine. It was a device; here's a conception of what this looked like. It was a globe filled with water, you would put heat under it until the water started to boil, and then there were these two nozzles extending in different directions. As the water boiled, the steam would expand, blow out the nozzles, and turn this thing. As you lit the fire under it eventually it would start spinning very rapidly with steam shooting out. It's kind of a little steam engine. Heron apparently did not see that this could be put to great use in all sorts of things, but at least he invented this. It was a kind of applied science.

He also showed people how to drill a tunnel from both sides of a mountain at once and meet in the middle. There's an application for you. How in the world do you do this? They actually did this in Classical times using Heron's directions. Nowadays we're not surprised when they dig the tunnel and it goes from Britain and it goes from France, and they meet exactly somewhere under the English Channel, but we're not talking about the modern world. Heron didn't have lasers and GPS systems or anything of the sort, but he had geometry and a brilliant mind, and he could explain how to dig such tunnels. As I said, they actually did this, so that's neat.

His great work is called the *Metrica* and this is where you find the two results I particularly want to tell you about. This means measurement, and so he was interested in measuring things, which of course is a good thing to do in applied mathematics. Interestingly, this complete work was only discovered by somebody named R. Schöne in Istanbul in a library in the year 1896, so that this is not something that had been around for thousands of years. It had been lost and then finally rediscovered just at the end of the 19th century. There had been references in other works about Greek mathematics that told us essentially what Heron had done, but it's interesting that the complete work took this line to be uncovered. It makes you wonder what other great works might be out there in some dusty library somewhere. In any case, Heron gave us the *Metrica* in which appeared his two works I want to talk about; his square root approximation, how to approximate square roots, and Heron's formula for triangular area, the star result, the main result of this lecture.

Let me start with his square root approximation. Suppose you want the $\sqrt{336}$. In fact, suppose I challenged you to find the square root of 336 and you immediately pull out your calculator, but no calculators. You might pull out your table of logarithms, no logarithms. I'll give you nothing but a stack of paper and a pencil, or maybe a stack of papyrus if we want to be in the spirit of Heron. How do you do something like this? How do you come out with a square root approximation without all these advances of logarithms or calculators or other things that we're so accustomed to? Heron showed how to do this. This is a really slick argument. Let me show you how to do the √336 à la Heron.

He says, first of all, you need an estimate. You need a first guess. Let me take 18 as my first guess for the √336. That's not right, but it's not a bad first guess because if you square 18, 18 × 18 is 324. That's not 336, but you're in the ballpark. That seems like a first guess, a good place to start. Then, Heron's technique shows you how to get a better guess from your first guess. He says as your second estimate use x_2 I will call it, my second estimate, as x_1, my first estimate, divided by 2 plus 336, the number whose square root we seek, divided by 2 times x_1. I'm taking my previous estimate over 2 plus 336 divided by the double of my previous estimate.

Let's try that. That would be 18/2, remember my first guess was 18, so 18/2 + 336/2 × 18, so that's 9, 18/2, and 336/2×18 is 336/26, and you can divide top and bottom by 12 and you get 28/3. What I would then decide is my next estimate is 9 + 28/3, if you get a common denominator there you get 55/3 as my second estimate, as a fraction—which of course is all the Greeks had to go on. If you wanted to break it off as to a whole number plus a piece, it turns out to be 18-1/3. There's your second estimate, 55/3 or 18-1/3. You've sort of refined the first estimate of 18 upward a little bit to 18-1/3.

Then Heron said let's get another estimate, a next estimate. What do you do? We'll call it x_3 and what I will do is take x_2, my previous estimate over 2, plus 336 over twice x_2—exactly the same process I just did to go from first estimate to second, I repeat that to go from second estimate to third. What's going to happen here? Remember if x_3 is $x_2/2$ + $336/2x_2$, we said x_2 was 55/3 from the previous one, so it's 55/3 over 2 + 336/2(55/3). The fractions here are a little bit nasty; you'd have to simplify these and get a common denominator. But, you're sitting there in the room with your big stack of papyrus, you can do this by hand, and when you do it turns out to be 6049/330. At that point I'm going to stop. I'm saying good enough. That's my estimate. If you break it off, you get 18 + 109/330, that's going to be my estimate of the √336. There it is—estimate √336 is roughly I think 6049/330. Then the question is—there are two questions—is this any good and, if it is, why does this work? What's going on here? Let me address both of these.

Is this approximation any good? Is the √336 somewhere near 6049/330? One way to see it would be to square 6049/330. After all, if that's the square root of 336 and I square it, I should get 336 or pretty close. If you square this, you get 36,590,401/108,900. If I divide this out, this turns out to be 336 plus a little more—how much, 1/108,900. That's awfully close to coming out right where I want it at 336. That's a heck of an approximation. But, if you don't like that, let's convert these to decimals. It's not something the Greeks could've done, but we can. Heron's estimate of 6049/330 converts out as a decimal 18.33030303 forever. If I get out my calculator and be very non-Heronian but be very modern and punch in the square root of 336, I get 18.33030278. If I round this to six places, they're the same. I am within six places in accuracy and it took me two steps in this approximation scheme with Heron. I guessed x_1 as 18, I refined it once, I refined it twice, and I got six-place accuracy. That's pretty impressive. That's Heron's scheme.

The question is why? Why does this work? Can you explain this? Can you justify this reasoning? Yes, you can. It's going to require me to delve into the modern notation of limits, but bear with me and you can see why this actually works. Remember what he did. He said x_1 was his first guess; x_2 was $x_1/2 + 336/2x_1$. Then x_3 was $x_2/2 + 336/2x_2$ and x_4, if I had gone further, would've been $x_3/2 + 336/2x_3$, and on and on and on. We just would continue this. The assertion is as you do this your approximations x_1, x_2, x_3, x_4 should be going to the √336 for some reason—for a reason I'm going to show you.

Here's what I'm going to say. Suppose as N gets bigger and bigger, we usually write it this way, as n goes to infinity, these x's are going to some limit which I will call L. If I look at x_1, then x_2, then x_3, then x_4 and see where they're headed, they're kind of converging to some limit L and I've got to figure out what L is. If Heron's right, L is the √336, that's what these things should be tending to, but that's not at all clear that's going to work. Let me show you why in fact L is the √336.

Let me put up on the screen here again the steps. There's x_2 is $x_1/2 + 336/2x_1$. There's x_3, there's x_4, and on we go. Remember what we said that as N gets bigger and bigger, these x_n's are approaching some limit L, whose identity I need to determine. What is L? Here's how you reason it. Let me look first of all at the column just to the right of the equal sign. You see that there's

$x_1/2$, $x_2/2$, $x_3/2$, and I want to know where are those heading? Where are they going? If x_1, x_2, x_3 are marching to L, $x_1/2$, $x_2/2$, $x_3/2$, etc., are marching to L/2. They're just half of the original term, so that limit would be L/2. I'm going to say that that column has to be going to L/2.

We're going to add to it the far right column. Where are those going? I've got $336/2x_1$, then below it $336/2x_2$, $336/2x_3$, where are those headed? If the x_1, x_2, x_3's are going to L, that column is moving toward 336/2L. It seems right. Let me put in the equals sign and let's look at the column on the left. What I see there is x_2, x_3, x_4, where are they going? The original group x_1, x_2, x_3, x_4 was going to L. If I sort of start at x_2, it's still going to L, x_2, x_3, x_4 are headed toward L. The limit on the left side is L. Thus I have this equation that has emerged by imagining what happens as you let N go to infinity. As you take more and more terms, where is this all headed? What I've concluded is it's headed to the expression that L, whatever that limit is, has to be L/2 + 336/2L.

The end is near. What I will do is take that equation L = L/2 + 336/2L and figure out what L is. Remember that was the object. What is this limit? Where are these x's going? You just solve this as an algebraic expression. I will bring the L/2 to the left. I'll now have L – L/2 is 336/2L. Of course on the left L – L/2 is L/2. Now I have L/2 is 336/2L, trying to solve this equation for L, cross multiply, and you'll get $2L^2$ is 2(336), trying to solve this for L. How about we cancel the 2's out of there? They're gone, so L^2 is 336. What's L? It's the $\sqrt{336}$. These x's have to be going to L and we just figured out what L has to be, the $\sqrt{336}$. Heron was right; this works and that's why. It's really quite a nice piece of mathematics.

If you want the general idea, suppose you want the $\sqrt{A}$. It doesn't have to be 336. You say xn is your approximation at any stage of the $\sqrt{A}$, then what Heron's technique says the next approximation will be we can write it as x_n + 1, the next one up is $xn/2 + A$. You always put the thing you're taking the square root of in the numerator over the double of your previous estimate, over $2xn$ and so the pattern that I have generated is just exactly this x_{n+1} is $x_n/2 + A/2x_n$. Sure enough as n goes to infinity the very same reasoning shows that those xn's will go to the $\sqrt{A}$. This is Heron's square root approximation scheme. Now you can come out of the room with your papyrus, you've got a way to figure out square roots by hand.

That's the one result that is in the *Metrica* that I wanted to mention, but the even greater one is Heron's formula for triangular area. You might say why do I need a formula for triangular area? We've already seen this. If you have a triangle with base b and height h, we saw Euclid gave us the formula that the area of the triangle is $1/2bh$. Why do I need another formula? It seems like I've got this one nailed down. No, no says Heron. What if you don't have h? What if you don't know the altitude of your triangle? For instance, suppose I had this triangle, 5 meters on one side, 8 meters on the other, 11 meters across the bottom, and I want to know its area, maybe for tax purposes. The tax collector is coming and I've got to know how much area this is. The formula that the area of a triangle is $1/2bh$ will do me no good at all here because I don't have the altitude. I need some other approach.

Yet, I observe this, there is an answer, there is a unique answer. There is only one area this triangle could have and that's because of the congruence. If you had another triangle that had a side 5, a side 8, and a side 11, those two triangles would be congruent by side, side, side, from Euclid Book 1, Proposition 8. What congruence means is that the triangles are identical, absolutely the same. They have the same angles, the same sides, and the same areas. Any other triangle with these dimensions has the very same area; the trouble is I don't know what it is. I've got to find it and that's what Heron's formula will give me.

What does Heron do? How does he give this area? He says let's suppose a triangle has sides a, b, and c, just arbitrary sides. I want to figure out the area knowing the three sides only. The first thing you do is you find what's called the semiperimeter. The perimeter of a triangle of sides a, b, and c is just $a + b + c$; they just go around. The semiperimeter is half of that. You take half the perimeter and you call that s, the semiperimeter, $a + b + c/2$, and then Heron's formula says the area of the triangle is, get ready, the $\sqrt{s(s-a)(s-b)(s-c)}$.

That can't be right. I remember the first time I saw this, I was in ninth grade, and we had been studying areas of triangles. We had just seen the area of the triangle was $1/2bh$ and there was a footnote and it said this. It said the area of a triangle can also be found as the $\sqrt{s(s-a)(s-b)(s-c)}$ and I said that must be a typo. That can't possibly be it. That's so complicated. The area of a triangle surely isn't that sophisticated, but in fact that was right. They didn't show me why it was true back in ninth grade, but I later saw that in fact this is right, this is Heron's formula, and it's quite amazing. It's my favorite result from geometry just because it's so implausible.

Let's go back to our triangle here and see how this works—triangle 5, 8, and 11. We're back in Heron's day and we've got to figure out the area of that for the tax collector who's coming around. What you do first is find the semiperimeter. That's half the perimeter, so $a + b + c$ is $5 + 8 + 11$, you divide that by 2, and you get 24/2 or 12. Half the perimeter is 12. Then, Heron's formula says the area of a triangle is the $\sqrt{s(s-a)(s-b)(s-c)}$: s is 12, the semiperimeter, a, b, and c are the three sides, 5, 8, and 11, so you substitute and you'll get, according to Heron, the area of the triangle is the $\sqrt{12(12-5)(12-8)(12-11)}$. You figure this out. This will be $12(12-5)$, which is 7, times $(12-8)$, which is 4, times $(12-11)$ is 1. You multiply all of this together and you see the area of the triangle is the $\sqrt{12 \times 7 \times 4 \times 1}$, which is the $\sqrt{336}$. Wouldn't it be nice if I could find the $\sqrt{336}$? Heron could. That was my previous example. This is why he needed to approximate square roots in the *Metrica* because he runs into them right here with the area of the triangle. Heron would get to work and show us rather quickly that the $\sqrt{336}$ is about 6049/330. That would be his area for the tax collectors' purposes. It's a wonderful, wonderful result.

Suppose I had a quadrilateral, a four-sided figure, which was 10 yards by 17 yards by 25 yards by 24 yards. I need the area of this. How do I find this? Actually, there is no unique answer here. For a quadrilateral knowing the four sides is not enough to nail down the area. I assume you know that, but if not just imagine a square, maybe $5 \times 5 \times 5 \times 5$, it has area 25. If you were to tilt it you'd still have $5 \times 5 \times 5 \times 5$ in a thin-looking diamond, which has much less area. Just knowing the sides of a four-sided figure does not determine the area. You need more information. With a three-sided figure, if you know the three sides, the areas are set by side, side, side. With four sides, all bets are off. If this was my plot of land and I walked around 10 yards and 17 yards and 25 yards and 24 yards, I'm not quite done. But, what I do then is you walk the diagonal. I march along that way and let's say that came out to be 26 yards. Now I have the four sides of my quadrilateral plus a diagonal. Now I can find the area and you do it by just applying Heron's formula to the triangle thus formed on the left whose area is A_1 and the one on the right whose area is A_2.

Let's figure this out. That triangle on the left has area A_1. I'm going to figure it out by Heron's formula. I first have to do the semiperimeter. There's 10 + 24 + 26/2. That comes out to be 60/2 is 30. The semiperimeter is 30 and hence A_1, the area of the triangle on the left, is the $\sqrt{s(s-a)(s-b)(s-c)}$. That'll be the $\sqrt{30}$, that was s, 30 – 10 is 20, 30 – 24 is 6, 30 – 26 is 4. I'm going to have to multiply 30 × 20 × 6 × 4, multiply all those together, and you get 14,400, take the square root of that, and I don't need Heron's approximation scheme because this one comes out perfectly, 120, exact. The area of the left-hand triangle is 120.

How about the area of the right-hand triangle? It's the same thing. We know three sides. We first have to figure out the; 17 + 25 + 26/2 turns out to be 34. Now you apply Heron's formula for the area, the $\sqrt{s}$, which is 34, times 34 – 17, which is 17, times 34 – 25, which is 9, times 34 – 26, which is 8. I've got to take the square root of that product. You multiply it all out, you get 41,616, and I don't need an approximation for that either because the square root of 41,616 is 204. How amazing that they came out evenly. I actually made it up that way. Now I'm done because the area of the quadrilateral is A_1 + A_2, A_1 was 120, A_2 is 204. Bingo—324 is the area of that four-sided figure.

You can imagine if you had a five-sided figure or a six-sided figure or any polygon, you just put in enough diagonals to break it up into some triangles and apply Heron's formula repeatedly, add them up, and you've got it. This is a very useful piece of mathematics.

I'm not going to prove this right now. We don't have much time and actually Heron's proof is rather complicated. It's a sophisticated geometry. I've shown you here the diagram that accompanies it. You can see the triangle there, a, b, c, he sticks a circle inside of it, and draws all these lines all over the place. It's a sophisticated, complicated piece of geometry. We're not going to prove Heron's formula here, but hang on. When we get to Isaac Newton, he had a proof of Heron's formula, much simpler, and that I will show you. You just need to wait a bit and you'll see Newton's proof of Heron's formula. Fortunately, it's not Newton's proof of Hero's formula, but we'll see that.

I'm not proving it, but I do want to just show you one last thing before we leave this great result. Heron's formula, if you accept this, if you have it, implies the Pythagorean theorem as a consequence. We've seen the Pythagorean theorem proved with the Chinese diagram with a little cockeyed square within a square and I mentioned that Euclid proved the Pythagorean theorem with the windmill. Here comes another derivation, this time using Heron's formula. I will be given a right triangle A, B, C, there it is. Let's say the sides are x, y, and z and I want to show that z^2 is $x^x + y^2$, the Pythagorean theorem I want to show follows.

Here's what I'll do. I will extend BC leftward to a point D, which is exactly y units long, so that from B to C is the same as from D to C. They both have length y. Then I'll draw AD, that line. I say that line AD will also have length z because I have two copies of my triangle left and right. Those two triangles are congruent, the one on the left, the one on the right by side angle side, so the third sides are both z. There's the triangle we're going to be looking at.

I want to first find the area of the big guy, triangle ABD, by the usual formula. The usual formula for area of triangle ABD is $1/2bh$. The base is $2y$ and the height is x. That's easy enough. The area of that triangle is $1/2(2y)x$ and a little bit of algebra there will reduce that to xy. The area of the triangle ABD is xy and we'll store that information for a minute and attack the area in a different fashion. There's the triangle again, triangle ABD.

Now let's find the area with Heron's formula. It's the same triangle, ABD, but a different attack, different approach. Here's Heron's formula. What I need to do first is find the semiperimeter. I have to take half of the perimeter around the outside of the big triangle ABD. If you look at the sides there I'm going to get a z, across the bottom I'm going to get $2y$, and I'm going to go back up and get z. I've got to take $z + 2y + z/2$. That's going to be $2y + 2z/2$. You divide out and you just get $z + y$. The semiperimeter of the big triangle is $z + y$. With that in mind, I can do Heron's formula. The area of the big triangle, there it is again the $\sqrt{s(s-a)(s-b)(s-c)}$—we just have to carefully insert all the variables here. S the semiperimeter is $z + y$. Then I've got to take that semiperimeter minus one of the sides, so how about $z + y - z$. Then I've got to take $z + y - 2y$ and then I've got to take $z + y - z$. That's the pieces of Heron's formula.

Let's simplify this. The first parentheses is $z + y$; I'll leave that alone. But, look at the second one, $z + y - z$; $z + y - z$, that's just a y. The next one $z + y - 2y$ is $z - y$ and the last one $z + y - z$ is a y again. The area of this triangle is the $\sqrt{(z + y) \times y \times (z - y) \times y}$. Multiply the y's together, you get a y^2, multiply the $z + y$, $z - y$ together, and you get $z^2 - y^2$. That's the area of the triangle with Heron. Now we cash in on all of this by equating the two results. Remember we found the area of the triangle your ordinary way. We got it to be xy. We found it with Heron and we got it to be the $\sqrt{(z^2 - y^2) \times y^2}$. Square both sides of this and simplify. The square on the left is $x^2 y^2$, the square on the right the square root disappears, and you just get $(z^2 - y^2) y^2$. I see something; cancel the y^2 and I have $x^2 = z^2 - y^2$, move the y^2 over, and you get $x^2 + y^2$ is z^2. There's the Pythagorean theorem proved as a consequence of Heron's formula.

With that, we're going to have to end our journey through Greek mathematics. It's kind of a sad departure. I love their work. I'm so impressed with what they could do, how deep they pushed mathematics, and how expertly they advanced the subject and with such limited tools, without many of the modern mathematical results that we take for granted. But, they did it. They left a legacy in mathematics that's unsurpassed, the Greeks.

Al-Khwārizmī and Islamic Mathematics
Lecture 9

[The quadratic formula] is the most famous and most useful formula in algebra. This will solve any quadratic equation for you.

The golden age of Islamic science, scholarship, and mathematics is usually set between the 8th and 13th centuries. During this time, Islam had grown from its origins and spread across North Africa into Spain and Sicily, then spread eastward to India. In the heart of this glorious civilization, Baghdad, was an intellectual center called the House of Wisdom. Islamic mathematicians at this time took three important directions in their work. First, they studied and translated the Greek texts, such as Euclid's *Elements* and the work of Ptolemy known to us as *The Almagest*. They also absorbed Indian mathematics, which included trigonometry, arithmetic, and the base-10 numeral system that we still use today. One example of Indian mathematics from the 7th century is Brahmagupta's formula for the area of a cyclic quadrilateral, which yields Heron's formula for triangular area that we saw in the last lecture. Finally, Islamic mathematicians were known in particular for their advances in solving equations, and the name that comes up most often in this regard is Muhammad Mūsā ibn Al-Khwārizmī (c. 780–850).

Al-Khwārizmī's origins are unclear, but he eventually gravitated to the House of Wisdom in Baghdad and became a major scholar there,

Al-Khwārizmī's important book was titled, in Arabic, *Hisab al-jabr w'al-muqābala*. The word *al-jabr* later became Latinized to "algebra," and Al-Khwārizmī's name itself was Latinized to "algorithm."

working in mathematics, geography, astronomy, and astrology. He wrote two important books, one of which was *On the Calculation with Hindu Numerals*. This work was translated into Latin around the year 1200 and served as the introduction of Hindu-Arabic numerals into Europe. Al-Khwārizmī's other important book was titled, in Arabic, *Hisab al-jabr w'al-muqābala*. The word *al-jabr* later became Latinized to "algebra," and Al-Khwārizmī's name itself was Latinized to "algorithm."

Through a process that he called restoring and comparing, Al-Khwārizmī solved first-degree linear equations and second-degree quadratic equations. He did this by transferring the equations into one of six forms. Al-Khwārizmī used six forms rather than the two we would use because negative numbers were not considered legitimate. For instance, we could solve the quadratic equation $3x^2 - 5x = -2$, but if negative numbers were not allowed, we would have to rewrite the equation as $3x^2 + 2 = 5x$. We could then refer to Al-Khwārizmī's chapter on squares and numbers equal to roots to find a solution. We see an example of Al-Khwārizmī in action while solving the following problem: "What must be the square which, when increased by 10 of its own roots, amounts to 39?" In modern notation, the problem is: $x^2 + 10x = 39$. Looking at this problem geometrically helps us understand Al-Khwārizmī's solution process and arrive at the result, which is 3. This technique is now referred to as completing the square. We can also apply this process to the generic quadratic equation. Here, we do algebraically what Al-Khwārizmī did geometrically, and we end up with the quadratic formula: $x = -b \pm \sqrt{b^2 - 4ac} / 2a$.

Other great Islamic mathematicians who came after Al-Khwārizmī include the number theorist Thābit ibn Qurra and the writer of the *Rubaiyat*, Omar Khayyam, who developed a geometric technique that proved useful for solving cubic equations. Islamic works eventually found their way back to Europe and helped to revive European mathematical learning in the 12th century, at the dawn of the Renaissance. We'll visit this period in the next lecture. ∎

Suggested Reading

Joseph, *The Crest of the Peacock*.

Katz, ed., *The Mathematics of Egypt, Mesopotamia, China, India, and Islam*.

Plofker, *Mathematics in India*.

1. Use Bhramagupta's formula to find the area of the cyclic quadrilateral shown.

2. Use al-Khwārizmī's recipe to find the number for which (in archaic language) "the square plus six of its roots is 40." In other words, complete the square to solve the quadratic equation $x^2 + 6x = 40$.

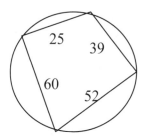

Al-Khwārizmī and Islamic Mathematics
Lecture 9—Transcript

We've taken our tour of Greek mathematics. We've met Thales and Pythagoras, Euclid, and Archimedes, and Heron. Now it's time to move beyond the Classical era to take a look at some of the mathematics of Islam and particularly the work of Al-Khwārizmī.

Let me give you some dates. In the year 476, Rome fell. This is the date traditionally attached to the end of the Roman Empire. In 529, Plato's academy in Athens was closed. This great center of learning for so many centuries was shut and this marks as well as anything the end of the Classical era of Greek and Roman scholarship. In the year 641, the Library at Alexandria burned. No one quite knows how this happened. There was a great conflagration and so much was lost. I think modern Classicists wish they could get in a time machine and go back to 640 and rescue all these documents because we just don't know what burned up in 641. This brings us to the Golden Age of Islamic science, scholarship, and mathematics usually set between the 8th and 13th centuries. This is the time period when Islam had grown from its origins and spread across North Africa into Spain and to Sicily and then spread eastward to India. In this glorious civilization there was an intellectual center at a place called House of Wisdom in Baghdad— Baghdad of course in modern day Iraq.

The House of Wisdom was like a great university. It was the successor to the Pythagorean brotherhood in southern Italy that we talked about to the Library at Alexandria in Egypt, now the center of mathematics moves to Baghdad to the "House of Wisdom." Islamic mathematics I will say went off in three important directions. Let me address each of these separately.

One thing they did was they studied and translated the great Greek texts. They had imported many of these back to Baghdad, they read them, turned them into Arabic, and found them as fascinating as we all do. They revered Euclid's *Elements*. They revered Archimedes and there was another text that they liked quite a lot, this was Ptolemy's work. We haven't mentioned him, but he was a Greek astronomer, Claudius Ptolemy, and he had written a great treatise on astronomy called the *Syntaxis Mathematica*, in Latin of course. This went back to Baghdad, was translated in the year 827, and the Arabic scientists thought this was quite impressive. It was in a sense the astronomical counterpart of Euclid's *Elements*. It was great. In fact they called it *Al-magiste*, the greatest. This was the greatest astronomical treatise they knew. When this Arabic translation goes back to Renaissance Europe and is turned back into Latin, they needed to translate *Al-magiste* and it became the *Almagest*. Nowadays we talk about Ptolemy's great work *The Almagest*, but Ptolemy never called it that. That's what happens when the language sort of goes around in a circle. *The Almagest* was a work preserved and studied by the Islamic mathematicians.

The second thing they did was absorb Indian mathematics. Remember the Islamic empire had extended all the way to the Indian subcontinent. They met the Indian mathematicians; there was a great tradition of Indian mathematics that had been flourishing. Islam was in well position because it was close to Europe and the Greek tradition, but also contacted the Indian one. The Indian mathematicians did trigonometry, they're often considered the founders of that subject, they did arithmetic and equation solving, and they gave us a wonderful numeral system.

Let me talk a bit about their mathematics and show you something called Brahmagupta's Formula, an Indian mathematician from the 7th century. This should ring a bell in some ways. Brahmagupta's looking at a quadrilateral, ABCD, a four-sided figure that's inscribed in a circle. Such a thing is called a cyclic quadrilateral, a quadrilateral in a circle. Here's a circle and there's quadrilateral ABCD. I should point out not every quadrilateral can be inscribed in a circle. If you imagine a diamond, a very long narrow four-sided figure, no circle can go through all four of those vertices, they're not properly positioned. Brahmagupta's talking about very special kinds of quadrilaterals, those that can be inscribed in a circle.

Let's say the length of the sides of this quadrilateral are a, b, c, and d. You go around, there's the sides. He wants to find the area of this inscribed quadrilateral, of this cyclic quadrilateral. First he's going to find the semiperimeter and this is where bells should be ringing. The semiperimeter is the perimeter $a + b + c + d$, 4 sides now over 2, and then Brahmagupta shows that the area of the cyclic quadrilateral is given by this formula, the $\sqrt{s(s-a)}$

$(s − b)(s − c)(s − d)$. That certainly harks back to Heron. In fact, you can think of Heron's formula as nothing but a degenerate case of Brahmagupta's, a special case of Brahmagupta's.

In this sense, let me show you how you can get Heron's result from Brahmagupta's. Suppose I'm given a triangle, triangle ABC. There it is. What I want to do is put a circle through the three vertices. I want to circumscribe a circle about this triangle. Can you do it? Yes, you can. How do we know? We know from Euclid. In Book 4, Proposition 5, Euclid shows you how to circumscribe a circle about any triangle. What you do is you bisect each of the three sides and then draw perpendiculars upward from there. They meet at a point; that point will be the center of your circumscribed circle. You make the distance from that point to any vertex be your radius, draw the circle, bingo, it goes through all three points. Any triangle can be put inside a circle, but not any quadrilateral can. There are the special quadrilaterals, the cyclic quadrilaterals, but it turns out that any triangle is a cyclic triangle.

I put a circle through these points. Now what I'm going to do is imagine the triangle ABC as if it were a quadrilateral ABCD, except D is the same point as C. Suppose it's got four vertices; it just happens that two of them land on top of each other, C = D. That's what it means by a degenerate case. The four sides degenerate down into three. Watch what happens to Brahmagupta now. The sides will have length a, b, c, and that extra side has length 0; d is 0 because c and d are in the same spot. One of the sides is length 0. When we go to the semiperimeter of the quadrilateral, which is really a triangle, we will add $a + b + c + d/2$, but d is nothing, 0, so it just amounts to adding $a + b + c/2$. In other words, the semiperimeter of the quadrilateral is just the semiperimeter of the triangle. Now you look at Brahmagupta's formula, which was that the area of the triangle when thought of as a quadrilateral, is the $\sqrt{(s − a)(s − b)(s − c)(s − d)}$, but $s − d$ is nothing but $s − 0$; in other words, it's just s. Brahmagupta tells me the triangle has area $(s − a)(s − b)$ $(s − c)s$ and that really should look familiar because that's Heron's formula. Brahmagupta's formula yields Heron's.

The Indians also gave us a fabulous numeral system. We still use it. The numerals in a base ten system appeared around the year 500. You find them on temples, carved on temple walls in India, and soon thereafter the 0 appeared.

You can look on one of these ancient temples and you see there's a 3, there's a 7, I recognize it, and there's a 0. The Islamic mathematicians, whenever they encountered the Indians, loved this base ten numeral system and brought it back with them to Baghdad and to the Islamic world. They called them the Hindu numbers, as indeed they were. Then, when these numbers got transferred back to Europe, the Europeans called them the Hindu-Arabic numeral system, which we still do, but in fact they're the Hindu numbers. The Arabs had just transferred them back to Europe.

The Islamic mathematicians studied and translated the Greek works and they studied and absorbed the Indian works. But, they were more than just conveyors of past work; they were also original creative mathematicians in their own right. In particular, the Islamic mathematicians are known for their advances in equation solving. The name that comes up most in this regard is Muhammad ibn Mūsā Al-Khwārizmī, whose dates are roughly 780–850. He was the great master of equation solving. Here's an image of him. Some artist imagined this is what he'd look like. Al-Khwārizmī's origins are not quite clear. He might've been Persian, he might've been Uzbek, but eventually he gravitates to the House of Wisdom in Baghdad and is a major scholar there, working not only in mathematics, but in geography, astronomy, and astrology. In those days, astronomy and astrology were sort of linked, unlike the present.

Al-Khwārizmī wrote two important books. One was *On the Calculation with Hindu Numerals*; that's the translation of it. When this was put into Latin, this is how these Hindu-Arabic numerals were introduced into Europe around the year 1200. From Al-Khwārizmī these ideas get back to Europe. His other great work has in its title the words Hisab al-jabr w'al-muqābala in Arabic. This is a work on equation solving. When this gets translated back into Latin, they had to do something with these words and in particular they had to make something out of al-jabr. How do we write this in Latin? How do we Latinize al-jabr? It turned into algebra. This word that is so pervasive in all of modern mathematics, algebra, is really the Latinization of part of the title of Al-Khwārizmī's equation-solving book. Actually Al-Khwārizmī himself got Latinized because his name whenever it was brought back and garbled a bit turned into algorithm. Whenever you talk about the Euclidean algorithm or computer algorithm, you are paying homage in a certain sense to Al-Khwārizmī. That's what happened to his name.

Al-Khwārizmī, through a process he called restoring and comparing, could solve linear first-degree and quadratic second-degree equations. He did this by transferring them into one of six forms. It seems to us, if I have to solve linear and quadratic equations, it's two forms, linear and quadratic. The Islamic mathematicians broke this into six forms because negative numbers, what we call negative numbers, were forbidden. They were not unknown. They were not considered legitimate. If you wanted to solve an equation, it could not have a negative in it. That required some gyration, some fancy footwork, to move things around and thus get six forms.

Let me show you the six forms. One type of equation Al-Khwārizmī could solve is what he called squares equal to roots. What's that? How would we write that? Squares mean something like ax^2, a bunch of squares. The root, the square root of x^2 is an x, so there's a bunch of x's. We would convert this squares equal roots to $ax^2 = bx$. If you wanted to solve an equation like that, you go to one chapter. Another chapter would be squares equal to numbers. This is approached differently. What's that going to be? We would write it as ax^2, some squares, equal c, a number. That's a different kind of equation. One is roots equal to numbers. Roots would be some x's equal to some numbers; we would write that as $bx = c$, a linear equation. One would be squares and roots equal to numbers. That would be $ax^2 + bx = c$, a quadratic equation. Squares and roots equals c, numbers, but you also better handle squares and numbers equal roots, $ax^2 + c$, numbers, equals roots bx. One more, roots and numbers equal to squares. That would be $bx + c = ax^2$. Six different treatments and we think that seems like an awful lot of extra work, but negative numbers could thus be avoided.

For instance, if we had to solve this quadratic, $3x^2 - 5x = -2$, we'd be fine with it, the negatives wouldn't bother us. But, if negatives are not allowed, you can't deal with this. This wouldn't have been something Al-Khwārizmī would've worked with. Rather you'd move things around and write it as $3x^2 + 2 = 5x$. There are no negatives in that and now you'd look up the chapter he has on squares and numbers equal to roots, which is what this is, some squares plus two numbers equals five roots, and you'd see how to solve it that way. This is another one of those strange surprises from the history of mathematics.

I want to give an example of Al-Khwārizmī in action while he actually solves a quadratic equation. I hope this will be illuminating because it's really quite an impressive piece of work. Here's the problem he gives. What must be the square which, when increased by 10 of its own roots, amounts to 39? It's expressed verbally. Remember, algebraic notation is still far in the future. Al-Khwārizmī is doing all of this verbally, even as the Greeks had to do their geometry verbally. This is his problem and here comes his solution. Here's what he says to do and I'll read this to you. It's a bit long, but it's illuminating. He said: You halve the number of roots, which in this case yields 5. Sure there were 10 roots; half of those is 5. This you multiply by itself; the product is 25. Add this to 39; the sum is 64. Now take the root of this, which is 8, and subtract from it half the number of roots, which is 5. The remainder is 3; this is the root of the square, which you sought.

Like so much of this old mathematics, you've got to dig through this to see what in the world he's doing. The best way to do that is to look at this geometrically. Here's the problem again. What must be the square which, when increased by 10 of its own roots, amounts to 39? We would write it this way: x^2, there's the square. I increase it by 10 of its own roots, so I add $10x$ to it, and this should be 39. You can see that what he's solving in our notation is $x^2 + 10x = 39$. How can we approach this as a geometry problem to see what he's up to? Let me take a square, and let's say it's x by x. Its area is x^2. The area of that square is the x^2. Remember what he says to do. You halve the number of roots, which in this case yields 5. We've got the 10 roots, the $10x$'s. What I'm going to do is sort of augment this square by going over 5 units and going down x units, and there's a blue rectangle there. It's 5 units across, x units down, so its area is $5x$. There's half the roots over there. Then, you do the same thing below. I drop down from the square 5 units, go across x units, and there's another $5x$. If you look at the picture as it now stands, there's a square and two rectangles tacked on the edges, but if you add up all those areas you've got $x^2 + 10x$'s; you split the $10x$'s into the two 5's. Since $x^2 + 10x$ is equal to 39, at the moment the area of this strange shape here is 39.

Now what? If you'll go back to Al-Khwārizmī's recipe, he says, this you multiply by itself—this meaning the number of roots, 5, you multiply it by itself and get 25. Look at the little lower right-hand corner of my picture. It's sort of missing a piece there; something's missing. What are the dimensions there? They are 5 by 5. If I were to fill that in with a green square, 5×5, I would've just added on 25, 5×5. What I'm going to do is, according to him, he says add this to 39 and the sum is 64. If I take the 25, the little green square, add it to the 39, which I started with, 39 and 25 is 64. We're doing some geometry here of some sophistication and where's it going. Let's look at the picture again. Remember, we're trying to solve $x^2 + 10x$ is 39. We have taken this square added on the two rectangular strips, put in the little green corner, and we know that that comes out to be 64. What's the area of this big square we've created? Look at it. It's $x + 5$ in each direction, so $(x + 5)^2$ is its area. That area has to be 64.

What does Al-Khwārizmī tell us to do next? He says take the root of this, which is 8—that is, he says take the square root of both sides. If $(x + 5)^2$ is 64, then the square root of $\sqrt{(x + 5)^2}$ is the $\sqrt{64}$. In other words, $x + 5$ is 8. Then he tells us to subtract from it half the number of roots, which is 5, the remainder is 3. He says take the 8, subtract half the number of roots, 5, and you get 3. That's what he claims the answer is, 3. This is the geometrical explanation of this kind of strange recipe he gave for solving his quadratic.

The proof of the pudding when you solve an equation is does it check—is x equal to 3? Let's see; you check this. What do you do? If I'm trying to solve $x^2 + 10x$ is 39, I put in 3. I get $3^2 + 10(3)$, and 3^2 is 9 and 10×3 is 30 and $30 + 9$ sure enough is 39, so it checks. The result comes out as advertised. This is right and this is the explanation for this. What he just did is what we call completing the square. You might've seen this in school, completing the square. Why is it called that? Look at the picture. That's exactly what he did. He filled in the little green square in the corner from which the solution of the quadratic equation became simple. The technique of completing the square we can trace back to Al-Khwārizmī.

We now would apply this to the generic quadratic equation. I want to just do that because I'm going to need to solve quadratics later in the course, so as we would think of this with all of our modern algebraic notation we would write $ax^2 + bx + c$ is 0. If there are negatives here, I don't care. Let's say we'll be very modern in our approach to this; we just want to make a general formula for the solution of the second degree, the quadratic equation. How do we do this? When we want to complete the square, we're taking our cue from Al-Khwārizmī. We're going to do the algebra. Here's what you do. You, first of all, divide everything by a. Instead of $ax^2 + bx + c$, you're going to have $x^2 + b/ax + c/a$ is 0. Then what I've done is move the c/a to the right, so it shows up there as $-c/a$. There's my equation, $x^2 + b/ax$ is equal to $-c/a$ and now you have to complete the square. What you're doing algebraically is what he was doing geometrically. I want to add something to both sides of this that'll turn the left side into a perfect square.

You can fiddle around with this for a while, but what you have to add to turn the left side into a perfect square is $b^2/4a^2$ and then that left side $x^2 + b/ax + b^2/4a^2$ will be the square of something. If I add it to the left, I better add the same thing to the right. Over there I will put $b^2/4a^2$ as well, and then if you look at that right side, you probably want to get a common denominator $-c/a + b^2/4a^2$, collect that all over $4a^2$, it will require you to turn the first fraction into $-4ac/4a^2 + b^2/4a^2$, turn it around, and you get $b^2 - 4ac/4a^2$, which might start to look familiar. The next thing you say is the left-hand expression, $x^2 + b/ax + b^2/4a^2$ is just the square of something. It is a perfect square. We have completed the square; it's actually the square of the quantity $x + b/2a$. The right side I don't do anything to, it's still $b^2 - 4ac/4a^2$.

In our road to solving this, I take the square root of each side. The square root of $(x + b/2a)^2$ is just $x + b/2a$. You just get rid of the square there. On the right, I've got to take the square root of the top and the square root of the bottom, the square root of the $4a^2$ I write as $2a$; the square root on the top, that's where I'll put the plus or minus. Again we realize we can use either plus or minus the square root to insert there. Let's do as they always do and put the plus or minus up there. I end up with plus or minus $b^2 - 4ac/2a$. That's $x + b/2a$. Finally, you take the $+ b/2a$ to the other side and you get x is $-b$ plus or minus the $\sqrt{b^2 - 4ac}/2a$, the famous quadratic formula.

This is the most famous and most useful formula in algebra. This will solve any quadratic equation for you. All you've got to do is spot the a, b, and c, stick the numbers in here. If there are two solutions, this'll find them. If there's one solution, this'll find it. If there are no real solutions, this will tell you that. In fact, this even works if you want to enter the realm of complex numbers. This is an all-purpose wonderful formula for solving second-degree equations. Is it from Al-Khwārizmī? Not exactly—he wouldn't have expressed it this way, but the strategy he employed of completing the square is exactly what you need in order to get this.

There are other great Islamic mathematicians of note who came after Al-Khwārizmī. One of them is Thābit ibn Qurra. I mention him because we'll see him again. He was a great number theorist. Later in the course we'll see a result that he stumbled upon and then it was forgotten. It was not transferred to Europe and people were retracing his steps many centuries later.

Another mathematician of note was Omar Khayyam, 1048–1131. Yes, this is the same Omar Khayyam who was the author. We know him for his work on the *Rubaiyat*. A famous author, people aren't so aware that he was also a very good mathematician. What he did was he tried to take a look not at second-degree equations as Al-Khwārizmī had done, but third-degree equations, the so-called cubic equations, to see if there was any way to solve cubics. He gave a geometrical technique that would provide some sort of idea of the solution of a third-degree equation. But, it wasn't as clean as an algebraic solution of the sort that I've shown you for the quadratic. Nevertheless, it was a step in the right direction toward solving third-degree equations.

Islamic works eventually found their way back to Europe through the Islamic empire, particularly in through Spain and Sicily, and they helped to revive mathematical learning in the 12th century in Europe. As the Renaissance is getting going, what's happened is these works have come back from the Arab world into Europe, filtered up into the early universities, and started the return toward the Renaissance, toward a knowledge base in Europe that had been lost for so many centuries.

Al-Khwārizmī's work in particular was translated into Latin; that's where the word algebra came from. In 1140, someone named Robert of Chester translated it, and just for good measure, in 1170 Gherardo of Cremona did. Al-Khwārizmī's work is available and it's a big hit in the mathematics of the time. People were reading this to learn about how to solve equations.

A math historian named Carl Boyer has surveyed the history of math and decided on the greatest textbooks of various eras. Here's his list. He says the greatest textbook of Classical times is Euclid's *Elements*. That was a no-brainer. That was easy. He said the greatest textbook of modern times is Euler's *Introductio in analysin infinitorum*. This is a great work. We'll talk about it when we get to Euler and this is from the 18th century. This is where Euler introduces functions, which become so important in modern mathematics. But, according to Boyer, the greatest work of medieval times, the greatest math text of that middle period is Al-Khwārizmī's *Algebra*. It's pretty hard to argue with that. This was a real masterpiece.

Where we're going from here is back to Renaissance Europe and back to one of the questions raised by Al-Khwārizmī's work. We can solve second-degree equations; how about third-degree equations? How about an algebraic solution for cubics? Is this possible? The Arabs couldn't do it. The Italian mathematicians of the 16th century in what is the weirdest story I know from the history of mathematics figured out how. That will be the subject of our next lecture.

A Horatio Algebra Story
Lecture 10

> It has been said that the 16th century produced three great scientific works. … We have a great work in medicine [Vesalius's treatise on the human body], we have a great work in astronomy [Copernicus's *De revolutionibus*], and the third major scientific work from the century is Cardano's *Ars Magna* of 1545.

B y the later 15th century, the Renaissance was up and running in Europe. The early universities were thriving at Padua and Bologna; with the advent of the printing press in 1450, books were widely available; and in 1492, Columbus discovered the New World. Europeans had absorbed the learning of the past and were now starting to push the frontiers of knowledge.

Our story in this lecture begins in 1494, when Luca Pacioli published the *Summa de Arithmetica*. This book described how to solve first- and second-degree equations but stated that a solution to the cubic equation (written in modern form as $ax^3 + bx^2 + cx + d = 0$) was impossible. The Europeans sought a "solution by radicals," defined as a formula that gives an *exact* solution and uses only the coefficients of the cubic equation and the "algebraic" operations of addition, subtraction, multiplication, division, and root extraction. Such a solution would be analogous to the quadratic formula.

Amazingly, Tartaglia independently discovered the solution of the depressed cubic and was able to solve all the problems given him by Fiore.

Early in the 16th century, a mathematician named Scipione del Ferro found a solution for the depressed cubic, a somewhat more restricted form than the general cubic. This equation is written as $x^3 + mx = n$; notice that the x^2 term is missing. On his deathbed, del Ferro passed this solution on to one of his students, Antonio Fiore. Fiore, in turn, used del Ferro's legacy to challenge the greatest mathematician in Italy at the time, Niccolo Fontana, nicknamed Tartaglia, the

"Stammerer." Amazingly, Tartaglia independently discovered the solution of the depressed cubic and was able to solve all the problems given him by Fiore.

At this point, a fellow named Gerolamo Cardano heard about the mathematical challenge and sought out Tartaglia to learn the solution to the depressed cubic. Cardano was a strange character, as we know from his autobiography, *De Vita Propria Liber*. He regularly conversed with his guardian angel and was a serious gambler but also became one of Europe's foremost physicians and wrote more than 100 books on a wide array of subjects. In 1539, Cardano promised Tartaglia that he would keep the solution to the depressed cubic a secret.

At around this time, Cardano began teaching mathematics to a brilliant young man named Ludovico Ferrari. In the course of their work together, Cardano discovered the solution to the general cubic equation, and his protégé discovered the solution to the quartic equation (written as $ax^4 + bx^3 + cx^2 + dx + e = 0$). But both of these solutions rested on the ability to solve the depressed cubic, which Cardano had promised Tartaglia he would not reveal. In 1543, Cardano went back to del Ferro's original papers and used those as the basis for publishing his great algebraic discoveries in the treatise *Ars Magna*.

Of course, Tartaglia was enraged and challenged Cardano to a mathematical contest. Cardano refused the challenge, but his protégé, Ferrari, accepted on his mentor's behalf. In the end, Ferrari bested Tartaglia, who disappeared into history, while the glory of this mathematical discovery is still attributed to Cardano and the *Ars Magna*. ∎

Suggested Reading

Cardan, *The Book of My Life*.

Hald, *A History of Probability and Statistics and Their Applications before 1750*.

Ore, *Cardano*.

1. The secret to solving a general third-degree equation rests in the special case of the "depressed cubic." This is a third-degree equation lacking its next-highest degree (i.e., second-degree) term—what Cardano would describe as "cube plus roots equals number" or we would write as $x^3 + mx = n$. Using this terminology, identify what a "depressed quadratic" would look like and indicate why it would be easy to solve.

2. We noted that Cardano's *Ars Magna* (1545) is placed alongside Vesalius's *De humani corporis fabrica* (1543) and Copernicus's *De revolutionibus* (1543) as the supreme scientific works of the 16th century. Why are the other two books so highly esteemed as scientific milestones?

A Horatio Algebra Story
Lecture 10—Transcript

We now move to Renaissance Italy. I want to tell the story of the solution of the cubic equation. In this lecture, I want to recount the tale and introduce the bizarre set of characters. In the next lecture, I want to show you the mathematics behind the algebraic solution of the cubic.

Here we go. In the 15th century, the Renaissance is up and running and Classical knowledge has been brought back to Europe. They're reading Aristotle; they're reading Al-Khwārizmī. The universities are thriving at Padua and Bologna. A particularly important date in the 15th century is 1450 when Gutenberg invents printing press. Now books are widely available. They can be published, read by far more people than were ever able to do so in the past. This is one of the great inventions of human history. Within three decades, the first mathematics book is printed on the printing press. Guess what it was? It was Euclid's *Elements* in 1482.

A bit later we have Columbus in 1492 discovering the New World. This was an amazing time. This is when the Europeans had absorbed the learning of the past and now we're starting to push the frontier. They weren't just reliant upon reading Aristotle or reading Plato. They could do things on their own. There was a sense of excitement and Columbus's discovery embodied that. They now knew there was a world across the Atlantic that Aristotle had not known of. Who knows what was possible. Who knows what sort of progress could happen.

My story begins in 1494 when Luca Pacioli published the *Summa de Arithmetica*, a mathematics book involving arithmetic and algebra. Luca Pacioli is famous not only for that, for his mathematical achievement, but he was the inventor of double entry bookkeeping, so he is well known among accounts everywhere. In his book, he describes how to solve the first-degree equations, second-degree equations, borrowing freely from Al-Khwārizmī, and then he gets to the third-degree equations, the cubic. He noted that the solution of the cubic equation was at that point impossible. Nobody could do it. In a sense, this was a pessimistic assessment of the situation, but perhaps also maybe a challenge to this coming century; let's see if we can push the frontier and solve the cubic.

What is the cubic equation? They would've said it's something like this: Cubes + squares + roots = number. We would write it more symbolically as $ax^3 + bx^2 + cx = d$ or actually in the modern version of this we'd bring everything to the one side and we would write $ax^3 + bx^2 + cx + d = 0$. For us, the a's, b's, c's, and d's could be positive or negative. The folks back in the 15th century were always insisting on positives, but let's say we'll go with that as the structure of the general cubic equation.

What was the challenge? What were they after? They wanted a "solution by radicals," it was called, or an "algebraic solution" of the cubic. Here's what that is. A solution by radicals is a formula, which gives the exact solution to the cubic—exact, not approximate—and uses only the coefficients of the cubic equation and the "algebraic" operations of addition, subtraction, multiplication, division, and root extraction. The formula that solves your equation has to be built from those components. Have we ever seen this before? Sure, we have seen it with the quadratic equation. In the last lecture, we looked at the quadratic equation $ax^2 + bx + c = 0$, second-degree. By completing the square, we cranked out the so-called quadratic formula. The solution to that quadratic equation famously is $x = (-b \pm \sqrt{b^2 - 4ac})/2a$. That is the algebraic solution of the quadratic. It's a solution by radicals because look at the formula. It has in it a, b, c, the coefficients, and they're added, subtracted, multiplied, divided, and square rooted. That's exactly what they were seeking for the third-degree equation. Could they find an analogous formula for cubics? That was the challenge.

Not too deep into the 16th century somebody did at least partially solve this. His name was Scipione del Ferro of Bologna. He found a solution for what's called the depressed cubic. It's not the general cubic, a somewhat more restricted form. He wrote it cube plus roots equals number. We would write it $x^3 + mx = n$, cube plus roots equals number. Notice what's missing here is the x-squared term. There is no quadratic term in this. That's why it's called depressed. It has nothing to do with its mood; it's just that it's lost its next highest term. It was this that del Ferro learned how to solve. He found the secret. We would imagine that he immediately told the world, right? No, he did not. He kept it to himself. To a modern scholar this is quite amazing. Here you have this extraordinary discovery, why don't you share it? But, there was a reason for this.

In Italy at this time, scholars could be challenged by other people to contests to prove their scholarly work. At any moment somebody could come out of the woodwork and challenge you. If you took them up on the challenge, there would be a public contest in which the two scholars would go head to head and whoever won might actually take the job from whoever lost. Your job was on the line and your reputation was on the line at all points. Therefore if you discovered something that nobody else knew, it was in your best interest to keep it to yourself rather than to share it with a wider audience because you never know when you might need it to meet a challenge, so del Ferro who has solved the cubic keeps it to himself.

He does this right up until his death at which point on his deathbed he imparts the secret to one of his students named Antonio Fiore. He passes the secret of the depressed cubic to Fiore. Fiore says, ah, I'm going to challenge the greatest mathematician in Italy at the time, whose name was Fontana. I'm going to challenge him and I'm going to use my knowledge of the depressed cubic to unseat him from his position to destroy his reputation. Fiore's going after this fellow, Niccolo Fontana of Brescia. Fontana had a nickname and it was "Tartaglia," the Stammerer as it translates. The reason for this was that as a boy Fontana had been attacked by French soldiers. Someone had slashed him as they raided his hometown right across the face destroying his palate, his tongue. He was not able to speak very well as a consequence. In fact, he almost died. It was said that only because a dog came along and licked his wounds did he survive. In any case, he acquired the nickname Tartaglia, the Stammerer. He was a very great mathematician and Fiore challenged him to this contest.

What happened was this: Each mathematician would give the other a list of problems. You'd have a certain amount of time to solve them and then in public you would reveal the solutions. Tartaglia gives Fiore a bunch of problems of various sorts. Fiore gives Tartaglia 30 problems and they're all the same. They are solutions of depressed cubics. Tartaglia looks at the challenge and he realizes he's either going to get none of these right or he too is going to have to discover the secret and he can get them all right. In what looks like sort of a Hollywood story, the clock is ticking and Tartaglia is hastening to try to figure out how to solve the depressed cubic in order to do this challenge. Amazingly, he does it. In fact, he tells us February 13, 1535, Tartaglia independently discovers the solution of the depressed cubic. It was the same solution that del Ferro had found and passed along to Fiore. With the solution in hand, Tartaglia gets all the problems right that he had been challenged with. His opponent Fiore doesn't do very well on his part. Tartaglia vanquishes his opponent; Tartaglia remains preeminent.

At this point, a fellow named Gerolamo Cardano learns about this challenge, hears about it, and says to himself, "Boy, I'd like to know how to solve the depressed cubic." Cardano, here's a picture of him, goes to see Tartaglia and begs him. Tell me the secret, Tartaglia. How do you solve the depressed cubic? Tartaglia is very skeptical about giving such a marvelous piece of mathematics—let me pause the story there, and jump back and talk a little bit about Cardano, introduce you to this person whom I regard as the most bizarre character in the whole history of mathematics, and then we'll pick up the tale in a minute.

We know a lot about Cardano because he wrote an autobiography, *De Vita Propria Liber*, the book of my life, in which he tells us about his strange and rich life. He tells us for instance that he was illegitimate, which in those days meant certain professions were restricted, his status was quite low, and he had to fight his way up to the top from those lowly beginnings. He tells us all about his health. This autobiography is full of a litany of his maladies; he tells us about his pimples, his boils, his fluxes, his digestion, and his impotence—more than we want to know. This is what they today call TMI, too much information, but it's just festooned throughout this book. He tells us about a way he had of feeling good. He said, "I considered that pleasure consisted in relief following severe pain…" Therefore, he said, "I have hit upon a plan of biting my lips, of twisting my fingers, of pinching the skin of the tender muscles of my left arm until the tears come," because when he then stopped, he had pleasure. The pain was gone. It sounds like that old joke about the person that hit himself in the head with a hammer because it felt so good when he stopped, except this wasn't a joke; this is how Cardano lived.

He believed in trinkets, amulets, and portents; he believed in guardian angels. He wrote this, he said: "Attendant or guardian spirits … are recorded as having favored certain men constantly—Socrates, Plotinus, Synesius, Dio, Flavius Josephus—and I include myself. All, to be sure, lived happily save Socrates and me …" It was one thing to believe in guardian angels, it was another thing to converse with them. In fact, Cardano would regularly be seen talking to his guardian angel and the public was somewhat suspicious of his sanity I think. In addition, he gambled, big time. In fact, he wrote the first treatise ever on probability called the *Book on Games of Chance*. In the whole history of probability this is where it started as he's giving the rules and the strategies for winning in various gambling situations. He gambled a lot. He said: "I gambled [at chess and dice] for many years; and not only every year, but—I say it with shame—every day." Also, he cheated. That got him into all sorts of scrapes.

He was quite a lusty character, but also a Renaissance man. He became one of Europe's foremost physicians. He was a doctor who was very good and his reputation extended well beyond his hometown all across Europe. For instance he treated the Pope. That was about as illustrious a patient as any physician could have. The Pope would call Cardano in for treatment and even more amazingly he treated the archbishop of Scotland. Somehow his reputation reached Scotland; the Scottish archbishop demanded that Cardano treat him and Cardano went to Scotland to do so. If you think about travel in the 16th century, that was a major journey and yet because of his fame he was called to that task. He also was a prolific writer on a wide variety of subjects. Cardano wrote on medicine; he wrote on mathematics; he wrote on religion; he wrote about astrology; he wrote about metoposcopy, which is telling the future by looking at the lines on people's faces, and many, many other things. He wrote well over 100 books in his lifetime on this wide array of subjects.

He had a son named Giambattista Cardano in whom he placed his hopes. Giambattista as he grew up was heading in his father's footsteps to become a physician and hopefully one with equal fame throughout the world. But, then, trouble strikes and Cardano tells us that on December 20, 1557, "… my bed suddenly seemed to tremble, and with it the whole bed-chamber." Things were shaking; things were moving. He took this to be a bad omen, one of these omens in which he put so much faith. Something bad was going to happen that day. What happened that day was he learned from a messenger that his son was going to get married. Cardano disapproved; the omen was against this. Something bad was going to come of this. He called the new wife a wild woman and that seems to be somewhat true because in their years of marriage, although she had children, Giambattista's wife told Giambattista that they weren't his.

The friction grew. Finally, Giambattista Cardano could take it no longer. Cardano's son baked his wife an arsenic cake. He fed it to her, she took a few bites, and died. Cardano's son murdered his wife for her infidelity. Giambattista Cardano was arrested, tried, convicted, and on April 10, 1560, Giambattista Cardano's beloved son was executed for murder. Cardano writes, "This was my supreme, my crowning misfortune." His life could go no lower than this, his beloved son executed as a common murderer.

It actually did go lower than that. In 1570, Cardano was jailed for heresy. He had cast horoscopes all his life, done astrology, but at this point he cast the horoscope of Jesus. This did not go over too well with the church and they threw him in jail, so he had committed sacrilege there I guess. It turns out somehow, in the remaining years he had, he rehabilitated himself. He got out of jail, he got in good with the Pope, and he lived with his beloved grandson, had a few happy years remaining in his life, and boasted that even then he had "...14 good teeth." Cardano, quite a guy, died finally on September 20, 1576, leaving behind quite a lot of good stories.

Back to the solution of the cubic. When we last left Cardano he was going to see Tartaglia begging for the secret, how do you solve the depressed cubic? Tartaglia resists, but Cardano was very persuasive. Eventually Cardano says I will make an oath. Here's what Cardano says to Tartaglia: " ... I swear to you by the sacred Gospel and on my faith as a gentleman, not only never to publish your discoveries, but I also promise and pledge my faith as a true Christian to put them down in cipher so that after my death no one shall be able to understand them." Please, if you give me this, I promise I'll never tell anyone. I just need to know says Cardano. Finally, Tartaglia relents. In 1539, he gives the secret of the depressed cubic to Cardano based on this promise of secrecy. Right away Tartaglia is regretting this, but he's given it away.

The story continues. A new character enters, Ludovico Ferrari. According to Cardano, one day the birds were chirping very nicely and he took this as a good omen. Something good's going to happen today. Who should show up at his door, but a youngster named Ludovico Ferrari and Cardano based on the omen takes Ferrari in. It turns out Ferrari is brilliant. Cardano starts teaching him things like mathematics. Pretty soon they are working together as a team, team Cardano if you will, working on great algebraic ideas. Cardano and Ferrari make a very powerful twosome.

Soon thereafter, Cardano discovers how to solve the general cubic equation. That would be one that does have the quadratic term in it. It's not depressed; it's general. It's what we have seen already as $ax^3 + bx^2 + cx + d = 0$. This was an advance. Then even more astounding, his young protégé Ferrari discovers how to solve the quartic equation, the fourth-degree equation, which we would write as $ax^4 + bx^3 + cx^2 + dx + e = 0$. Algebra was really expanding here. These were two great advances; however, here was the snag. Both of these solutions rested upon the ability to solve the depressed cubic. What Cardano would do to solve the general cubic was to reduce it to the depressed cubic and what Ferrari would do to solve the general quartic would be to reduce it to a cubic and then that would require the depressed cubic, but the depressed cubic was the very thing that Cardano had gotten from Tartaglia and he was not allowed to share with the world. He was not allowed to publish it, so it was a roadblock to his publishing any of these great discoveries.

What did he do? Remember the story, how this all unfurled. First, del Ferro had discovered the solution of the depressed cubic and given it to Fiore. Then, Tartaglia, from the necessity of the contest with Fiore, had discovered it himself and had given it to Cardano. Cardano had promised not to reveal Tartaglia's secret, but he had never made such a promise to del Ferro who was long dead, and so in 1543 Cardano went to the family of del Ferro, the original discoverer, saw del Ferro's papers, and saw how del Ferro had solved the depressed cubic. The fact is it was the same way that Tartaglia had, but Cardano wasn't going to reveal Tartaglia's secret. He was going to reveal del Ferro's and that wasn't a secret at all. Cardano writes up the great algebraic discoveries and publishes them in his book the *Ars Magna* in 1545. This is where the world learns how to solve cubic equations, the *Ars Magna*.

The book begins with this opening sentence. Cardano acknowledges there that, "This art originated with Mahomet..." *Ars Magna* of course translates to the great art. What was the art this book was about? The art was algebra. Who was Mahomet he's referring to? The art originated with Mahomet; that was Muhammad Mūsā ibn Al-Khwārizmī whom we have met as the founder of algebra. Cardano begins by giving credit to Al-Khwārizmī. The first few chapters of the *Ars Magna* are handling the first-degree equations and the second-degree equations that are essentially rehashes of what Al-Khwārizmī had done.

Then we get to chapter 11, the big one, where something new hits the stage. It's titled "On the Cube and Roots Equal to the Number," which we would write $x^3 + mx = n$. This is where he tells the world how to solve the depressed cubic. He begins this chapter with a little preface; this is kind of interesting. He says: "Scipio Ferro of Bologna well-nigh 30 years ago discovered this rule and handed it on to Antonio Fiore of Venice, whose contest with Niccolo Tartaglia of Brescia gave Niccolo occasion to discover it. He, Tartaglia, gave it to me in response to my entreaties, although withholding the demonstration." Cardano was upfront about it there that he didn't claim that he had discovered this for himself. He did give credit to Tartaglia in this little preface and kind of recounted the amazing tale of the discovery of the solution to the depressed cubic.

Tartaglia read this *Ars Magna* and went ballistic. He said Cardano you had promised as a gentleman and on your faith as a true Christian not to reveal my secret and here it is in print in the *Ars Magna*. Cardano would've responded I wasn't revealing your secret, it was that other guy's secret even though it was the same secret, but Tartaglia wasn't having it, and so he challenged Cardano to a contest. Here we go again, another contest. Cardano declined. He did not want to engage in a contest, but his protégé Ferrari said, "I'll take up the challenge for my mentor." Ferrari accepted the challenge from Tartaglia. Ferrari was brilliant, but he was also one tough cookie. He was missing fingers on one of his hands because of a duel in which he had been and he was always getting into fights. When he and Tartaglia met for their public challenge, Ferrari was vicious in his comments against Tartaglia.

His mathematics was better than Tartaglia's and he vanquished and thrashed Tartaglia in the contest. It is said that Tartaglia was lucky to escape with his life. Tartaglia runs away and sort of disappears into history, as all the fame and glory resides with Cardano and the *Ars Magna*.

What is this rule? What was the great discovery? How do you solve the depressed cubic $x^3 + mx = n$? Let me show you what Cardano wrote. This was the statement that he gives in chapter 11 of the *Ars Magna*. I'll read it, it's rather convoluted, but we'll see this again in more algebraic form, but just to give a sense of how complicated this is and how difficult it was to work in a presymbolic algebraic world. Here's how you do it. According to Cardano, you cube one-third the coefficient of x; add to it the square of one-half the constant; and take the square root of the whole. You will duplicate this, and to one of the two you add one-half the number you already squared, and from the other you subtract one-half of the same, and then subtracting the cube root of the first from the cube root of the second, the remainder which is left is the value of x. We can make no sense of this in this form, but trust me this is right. We'll see that in the next lecture.

It has been said that the 16th century produced three great scientific works. One of them was Vesalius's treatise on the human body in 1543 that sort of opens up the realm of medicine. One of them was Copernicus's *De Revolutionibus* in 1543 also—same year interestingly—which of course gives us the Copernican system of astronomy placing the sun at the center with the earth traveling around it. We have a great work in medicine, we have a great work in astronomy, and the third major scientific work from the century is Cardano's *Ars Magna* of 1545. Cardano lands on the list of the great masterpieces and notice the three dates there are almost exactly the same in the 1540s; it's kind of interesting.

Cardano was very proud of the *Ars Magna*. This was one of his great works. He was like a proud parent of this thing. At the end, the very last line of the *Ars Magna* referring to the book, he writes this rather touching sentiment. He said, "Written in five years, may it last as many thousands." The jury's still out whether it will last 5,000 years, but we're still looking at now in this course, so it's still around.

What I want to do in the next lecture is see what all the fuss is about. What was this solution of the cubic equation by a formula? How do you do it? What was his discovery? I think you'll see this is really one mathematical gem.

To the Cubic and Beyond
Lecture 11

You can sort of see at this point why they don't teach [the depressed cubic formula] in schools. It's hard enough to remember the quadratic formula, let alone this one.

In this lecture, we'll see how to solve the cubic equation, but we'll warm up by looking at a second-degree equation. As we've seen, the quadratic formula allows us to solve any second-degree equation of the form $ax^2 + bx + c = 0$. All we have to do is substitute the coefficients a, b, and c into the formula $x = (-b \pm \sqrt{b^2 - 4ac})/2a$. This is an analog of what we're seeking for the third-degree equation.

To solve the cubic equation, we will follow Cardano in *Ars Magna*. We begin with the depressed cubic, written as $x^3 + mx = n$, but we'll substitute numbers into the equation: $x^3 + 24x = 56$. Cardano looked at this problem in much the same way that Al-Khwārizmī looked at the second-degree equation. Instead of completing the square, however, Cardano approached the problem by subdividing a cube. He then had to find the volume of the cube by adding up the pieces. When we go through this process, we're left with the expression $u^3 + (t - u)^3 + (t - u)[2tu + u^2 + u(t - u)]$, in which t represents the height, length, and depth of the cube and u represents units. We set this expression equal to t^3, the volume of a cube. Simplifying, we get $(t - u)^3 + [3tu](t - u) = t^3 - u^3$, which has the same structure as the equation we're trying to solve, $x^3 + 24x = 56$. We substitute x for $t - u$, and Cardano's equation becomes $x^3 + [3tu]x = t^3 - u^3$. We now designate the coefficient $3tu$ to be 24 and $t^3 - u^3$ to be 56.

This leaves us with two equations in two unknowns, t and u. We approach these by solving one equation for one variable and substituting that back into the other equation. Here, we end up with $t^3 - 512/t^3 = 56$, leaving us with one equation and one unknown, t. To eliminate the t^3 in the denominator, we multiply both sides by t^3, which yields $t^6 - 56(t^3) - 512 = 0$. This is a six-degree equation in t, but it is second degree in t^3. It can be written as $t^6 - 56(t)^2 - 512 = 0$, which looks just like a quadratic equation. If we let y

play the role of t^3, we get the exact second-degree equation we saw at the beginning of this lecture: $y^2 - 56y - 512 = 0$. For y, or t^3, we then choose one of the two solutions we found earlier, 64 or -8, and work our way back up the chain to a solution. Using 64, we find that the solution to our earlier cubic, $x^3 + 24x = 56$, works out to be $x = t - u = 4 - 2 = 2$.

For the generic depressed cubic, the formula is: $\sqrt[3]{\frac{n}{2} + \sqrt{\frac{n^2}{4} + \frac{m^3}{27}}} - \sqrt[3]{\frac{n}{2}\sqrt{\frac{n^2}{4} + \frac{m^3}{27}}}$. We can try this formula with the example done in the lecture, substituting 24 for m and 56 for n. What if the cubic isn't depressed, as in the generic $ax^3 + bx^2 + cx + d = 0$? Cardano discovered that he could make the equation depressed by replacing x with $y - b/3a$, where the a's and b's are the coefficients of the general cubic.

It has since been shown that a solution by radicals of the general quintic or any higher-degree equation is impossible.

Cardano's *Ars Magna* also includes Ferrari's solution of the quartic, which he accomplished by reducing the fourth-degree equation to a third degree, then to a depressed cubic. It has since been shown that a solution by radicals of the general quintic or any higher-degree equation is impossible. ∎

Suggested Reading

Cardano, *Ars Magna*.

Ore, *Cardano*.

Questions to Consider

1. Use Cardano's formula to get a solution by radicals for the depressed cubic $x^3 + 15x = 124$. Check your answer.

2. Solve the cubic equation $x^3 + 9x^2 + 42x = 52$ by first making the substitution $x = y - 3$ to reduce this to a depressed cubic in y. HINT: The depressed cubic you get should look familiar.

To the Cubic and Beyond
Lecture 11—Transcript

In the previous lecture, I told the story of the solution of the cubic and we met quite a cast of strange, but great thinkers. This lecture is the great theorem that they were working on. How is it that you solve a third-degree equation by a formula?

I should say at the outset that there's a lot of math in this lecture. The whole lecture will be devoted to solving the cubic. It'll be dripping with symbols. There are probably more formulas per square foot than in any other lecture in the course. I'm going to try to go slowly and spell it all out; some people might want to have a pad and paper by their sides to follow along with it. It's challenging, but I think it's worth it because, a, this is a brilliant piece of work—this is mathematical artistry—and, b, you might never have seen it done. A lot of people aren't aware that there's a formula to solve third-degree equations. We'll see it; we'll see where it comes from. We'll follow Cardano in his description of this.

Before we start, given how challenging this is going to be, we need a little warm-up here, so let me warm-up by looking at a second-degree equation. Remember we have seen the quadratic formula, which allows us to solve any second-degree equation of the form $ax^2 + bx + c = 0$. All you've got to do is substitute the coefficients a, b, and c into that marvelous formula that says $x = (-b \pm \sqrt{b^2 - 4ac})/2a$. It is an analog of that that we're seeking for the third-degree equation. Let me actually solve a particular quadratic. Here it comes just to make sure we see how this all works. There's one: $x^2 - 56x - 512 = 0$. We want to solve this with the formula. I'm also going to let you in on a secret. I'm going to need this a little later in the lecture. This isn't just an arbitrary quadratic. I have a reason for looking at this particular one, as you'll see in a few minutes. Anyway there we go: $x^2 - 56x - 512 = 0$. We want to solve this with the quadratic formula.

I look at the formula. I know x is $-b$. Remember the a, b, and c are the coefficients, so in my quadratic the a is 1 because there's one x^2, the b is -56, that's how many x's there are, and the c is -512. I just have to put those in there. So x is $-b$ $-(-56) \pm \sqrt{(-56)^2} - 4$ times a, which is 1, times c, which is

–512, all of this over $2a$, which is $2(1)$. You'd start doing the simplifications. The $-(-56)$ is $56 \pm \sqrt{}$ now $(-56)^2$ is 3136 and the $-4(1)(512)$ is 2048, so that's under the radical. All of this is over 2. You add up the terms under the radical and you get 5184. Then when you take the square root of that you'll find that x is $56 \pm \sqrt{5184}/2$, but the $\sqrt{5184}$ is 72 exactly. You get $56 \pm 72/2$. When you add them, you get $128/2$. When you subtract them, you get $-16/2$. After all of this you get that x is either 64, $128/2$, or -8. There are actually two solutions to this quadratic. As I said, keep these in mind; we'll see them again.

That's the quadratic; how about the cubic? How about the third-degree equation? We're about to plunge in. Let me just say that as you might expect this is going to be harder. It's more sophisticated, but it's kind of incredible to see how this is done. Remember what we're doing. We're going to follow Cardano in chapter 11 of *Ars Magna* in which he first addresses just the depressed cubic, the sort of restricted sort, $x^3 + mx = n$. Rather than do that general cubic with an m and an n in it, how about just to make things a little easier I will put specific numbers in those slots. Instead of $x^3 + mx = n$, how about we'll pick this one, $x^3 + 24x = 56$? It's not going to be quite the generic general solution, but all the steps I do here can be replicated with m's and n's; it just makes it even more complicated looking, so I think this one with numbers in will make it a much more accessible kind of derivation. We're going to go after just that one, $x^3 + 24x = 56$.

This is a cubic equation. Cardano interestingly begins with a cube, an actual 3-dimensional cube. Remember when we saw Al-Khwārizmī solving a second-degree equation, he completed the square and he was looking at a 2-dimensional square. You go up to a third-degree equation, you go up a dimension, and so Cardano's looking at a literal cube and let's say it's $t \times t \times t$. I've drawn a picture there of a cube t units high, t units across, and t units deep. The volume of this is what we're going to try to track down. The volume of a $t \times t \times t$ cubed is obviously just t^3. That's easy enough. Cardano has to do something more sophisticated than that and he sure does when he subdivides the cube in a kind of interesting and, it turns out to be, very valuable fashion.

Here's a picture of my subdivided cube. What I'm doing is slicing this cube up into various pieces. It's the same cube, but it's now been broken up into the pieces you see. The cube is still t units high, t units across the bottom, and t units deep. That's the same cube, but you see down in the lower front corner, there's a little piece that's shaded. That's a little cube that's u units by u units by u units. We're going to have to worry about what the t and the u are. We'll get to that in a bit, but let's just suppose that that little cube down there is $u \times u \times u$. If the whole base of the cube is t units and I put that little chunk there that's u, that means the other piece which is labeled AB in the diagram would have to be $t - u$ units. Likewise going back on the right side, if I have a little chunk of u units, the remainder of that which is labeled DE is $t - u$ units. These are the dimensions of this subdivision of this cube.

What Cardano has to do is find the volume of the cube again by adding up all the little pieces into which it has been broken. We're going to go after the volume of the cube once more, but this time by collecting all the little blocks and pieces you see before you. You have to do this methodically; here we go.

First of all, let me get that little cube in the lower front corner. What's its volume? It's $u \times u \times u$, so that's u^3. Let me next get that big shaded cube in the upper corner. You can see it there; it's a little darker. We're going to get its volume. It's $t - u$ one way by $t - u$ the other way by $t - u$ in the third dimension, so it's a cube whose volume is $t - u$, the quantity cubed. What else do we have here? Look at the front; right here facing front there's this slab standing there. The base is AB, the height is the whole height of the cube, and its u units thick. It's facing us in the front, it's this big slab, its volume would be $t - u(t)(u)$, the product of the three dimensions. Let me put it in the form $tu(t - u)$, so that's the front facing slab. But, there's another slab on the right side that is exactly the same dimensions. It goes along the base DE and it rises all the way to the top, so it's $t - u$ by t and u thickness. I'm going to add that volume in. That's exactly what I just did, so I'm going to get another $tu(t - u)$, let's just put a 2 there and make this as $2tu(t - u)$.

I'm still not done. There are still some pieces left. Where are they? Look on the front here. There's that little shaded cube in the lower front corner, standing on top of it is a block. I don't have that yet. Its dimensions are u by u by $t - u$, so it's volume is $u^2(t - u)$. There's one more piece. If you look at that shaded cube in the upper back, it's resting upon a pedestal there below it. I need that volume and if you look at the dimensions there, it's $t - u$ by t

$-u$ by u, so the volume there would be $u(t-u)^2$. I think I've collected all the pieces now. There they are.

Now what we want to do is a little algebra on this. Here's what I'll do; I'll write down the u^3 that started it and the $(t-u)^3$. Leave those alone, but the next three terms, the $2tu(t-u)$ plus the $u^2(t-u)$ plus the $u(t-u)^2$, notice each of those has in it the $(t-u)$ expression. It's common to all three of those last terms, so let me factor out $t-u$ from those three. When I do that, the first term will remain $2tu$, which I'll put in the square brackets. When I pull the $t-u$ out of the second term, I'll be left with a u^2 in the square brackets. When I pull a $t-u$ out of the third term, I'm left with a $u(t-u)$. This expression on the screen is the volume of the cube, but we also know the volume of the cube quite simply is t^3 and so I'll set those equal. We've done the volume of the cube both as one big cube and as all these little pieces.

For some reason, this is going to get us somewhere. It's still not at all clear where this is going. Here's the equation: $u^3 + (t-u)^3 + (t-u)[2tu + u^2 + u(t-u)] = t^3$. What are we going to do with this? Here's what we'll do. On the right side is that t^3; on the left side it starts off with a u^3. Let me move the u^3 over to the other side. The u^3 goes to the right and now on the right I have $t^3 - u^3$. In terms of what's left on the left side, it starts with that $(t-u)^3$ and then there's this next expression, but look in the square brackets there. I see in the midst of that $u(t-u)$; let me multiply that out. That'll give me plus $t-u$ times [the $2tu$ plus the u^2 plus another tu, when the u hits the t, and then minus a u^2, when the u hits the minus u. That seems to cry out for something to be done in those square brackets. I have a u^2, I have a $-u^2$, so they cancel. Now we're getting things a little simpler because inside the square brackets there's a $2tu$ plus a tu, that's $3tu$, and that's all that's there. I'm finally reaching a point where I can take a deep breath here. Let me clean this up and say that what I've got then is the $(t-u)^3$, the first expression, plus the square bracket term [$3tu$] times its multiplier $(t-u)$ is equal to on the right side $t^3 - u^3$. That's important. I'm going to call that star; I'm going to refer to that in a minute.

What is the point of all of this? Here's the critical thing. This is the sort of thing that Cardano would've seen. That has the structure of the equation we're trying to solve. Remember we're trying to solve $x^3 + 24x$ is 56. Something cubed plus 24 times the something equals 56. If you look at equation star, you see a mirror image of that, if I were to let $t - u$ be x. In the equation star, take out the $t - u$, put in the x, and star will become $x^3 + [3tu]x = t^3 - u^3$. It has that same flavor as a depressed cubic. But, the cubic we really wanted was $x^3 + 24x$ is 56. That's what we're trying to solve. Now I look at these two depressed cubics; they both have an x^3 in them. In one case I have $3tu$ times x, below I have 24 times x. On the right I have $t^3 - u^3$ in general, but the 56 in the specific case suggests I should equate these. Let the coefficient $3tu$ be 24; let $t^3 - u^3$ be 56.

What I have now done is found two equations in two unknowns, t and u. If you remember your high school algebra you can solve two equations in two unknowns. That's what I'm going to have to do. Suppose I could do that. Suppose I could figure out what t is and what u is. Then up above I said x is $t - u$, I can subtract those, and I would have figured out what x is. That's the strategy; that's where we're headed. I now have those two equations, $3tu$ is 24 and $t^3 - u^3$ is 56, and I've got to solve these simultaneously. The way you do that is you solve one equation for one letter and substitute that back into the other. Let's look at that top equation: $3tu$ is 24. That tells me that u, let's solve that for u, is 24 over $3t$. Just bring the $3t$ underneath on the right, but 24 over 3 is 8, so this is just 8 over t. That's what u is. U is $8/t$. Go back to the second equation, $t^3 - u^3$ is 56, u is $8/t$, stick it in there. You'll get $t^3 - (8/t)^3$ is 56. If you cube that $8/t$, you'll end up with $t^3 - 512/t^3$ is 56.

This is a landmark on the road to the solution. We're not quite there yet, but notice what I've got now—one equation, one unknown, t. If I can figure out what t is, I can go back and figure out what u is. I can go back up the line and figure out what x is, which is what the object is here. Now we just have to solve that equation. That's still not quite there yet. One of the troubles is it's got a t^3 in the bottom. You don't want a variable in the denominator, so I'm going to multiply both sides of this by t^3. When I distribute across, I can cancel out that t^3 on the bottom. When I do that I get $t^3(t^3)$ is t^6, I get $t^3 - 512/t^3$ is just -512 and on the right side I get $56t^3$. One more thing—let me move everything over to the left and so here's the equation I've got to do. This is what it's all going to come down to. Can I solve t^6...that's what $t^3(t^3)$ will be t^6, there'll be six t's...$- 56(t^3) - 512$ is 0. That's the ballgame. Can we solve this?

My first thought is this is worse. If you think about it, we started trying to solve a cubic, a third-degree equation—that's a six-degree equation. It seems like I've made life worse. It's a six-degree equation in t. But—and here's the real thing, here's the key to unlocking the whole problem—yes, it's six-degree in t, but it's second-degree in t^3. It's quadratic in t^3, by which I mean I can write that as the quantity $(t^3)^2$ (that's t^6) $- 56(t^3) - 512 = 0$. That looks like a quadratic equation. In fact, it looks just like $y^2 - 56y - 512$ is 0 if you let y play the role of t^3. I've reduced the problem to solving that second-degree equation, but that's the exact second-degree equation I solved at the beginning of this lecture. I had an x instead of a y, but it was the same one. If you look back, you see that we used the quadratic formula to figure out what y is. Now, y is t^3, but numerically y came out to be 64 or –8. Using the quadratic formula in the course of solving the third-degree equation, I decided that t^3 is 64 or –8. You only need one of these two solutions, so how about we use the t^3 is 64.

Now all we've got to do is back up the chain and we can do it. We're done. If t^3 is 64, then t is the cube root of 64, which is 4. Now we've nailed that down; I know what t is. I better figure out what u is. Remember we saw somewhere up the line that u was $24/3t$, which we reduced to $8/t$, but I now know that t is 4 so u is 8/4, u is 2. Now the dramatic conclusion, x is what we wanted, the solution of our cubic, x is $t - u$. I've just said that t is 4, u is 2, so x, the solution of the cubic that we started with, $x^3 + 24x$ is 56 would be $x = t - u = 4 - 2 = 2$. That is how you do it. That's the solution of the depressed cubic. It's quite a journey, quite a workout, but it's kind of neat. The first thing I better do here before we proceed is check it. Is this right? I better be sure. You always want to check your answer: x is 2 is my candidate for the solution, so if x is 2 I'll stick it into $x^3 + 24x$. That'll be $2^3 + 24(2)$, 8 + 48, 56, yes. That's encouraging; it checks.

What you want to do in general is look at the depressed cubic $x^3 + mx = n$. This is the generic cubic; it doesn't have the 24 and the 56 in it. This is any cubic, any depressed cubic, and you do the same thing. You take the cube t by t by t, you break it up into little pieces, and you get those equations spinning out. You let x be $t - u$, you solve for t, you solve for u, and eventually you get a general formula, the cubic equation, and I will show it to you. This is what comes out when you do it with the m's and n's. X turns out to be this giant formula. The cube root of $n/2$ plus the square root of $n^2/4$ plus $m^3/27$ minus the cube root of $n/2$ plus the square root of $n^2/4$ plus $m^3/27$. In my previous lecture, I read to you Cardano's solution for the depressed cubic. It was a big paragraph with lots of complicated words in it. If you translate that verbal description into symbols, you get exactly this. This is what Cardano gave as his solution. Notice one other thing. This is a solution by radicals. It's just what we wanted because all I need to know is m and n, the coefficients of my cubic. I stick them into this formula, which involves the m and the n added, subtracted, multiplied, divided, square rooted, and cube rooted, all of which are allowable algebraic operations and I'll get the solution.

Let me check this in this sense. Let me return to the example I just did, $x^3 + 24x = 56$, and try this formula out on that particular example. I know what the answer should be. It should be 2. We did it. We derived it from scratch. Let me now just try to derive it from the formula. I would first identify that m is 24. Remember the pattern is $x^3 + mx$ is n. The m is the 24 and the n is the 56, and I just have to stick those numbers into this massive looking equation. I'm going to do that, but I see one way to streamline this a little bit. If you look at the solution for x, there's a cube root in the midst of which there's a square root of $n^2/4 + m^3/27$. Then there's another cube root subtracted from it, but it has that same square root underneath. Maybe I'll just do that square root all by itself and then we'll go back and plug in the numbers.

What I will say is let's first just figure out the square root of $n^2/4 + m^3/27$. N is 56, so I put in the square root of $56^2/4$. M was 24 so I add to it $24^3/27$. You multiply this out and you get the square root of $784 + 512$, which comes out to be the $\sqrt{1296}$, which is a perfect square, outcome's 36. In the solution where I see the $\sqrt{n^2/4} + m^3/27$, in both spots I'm just going to stick in 36. Now I can quickly run to the end. I'm trying to solve $x^3 + 24x$ is 56. I've got my cubic formula. I put in the numbers, I get the cube root of $n/2$, which is 56/2, plus that square root I just did, which we said turned out to be 36, minus the cube root of $-n/2$, $-56/2$, plus that same square root 36. Since 56/2 is 28, I'm now down to the cube root of $28 + 36$ minus the cube root of $-28 + 36$; 28 and 36 is 64, so I've got the cube root of 64 minus $28 + 36$ is 8, minus the cube root of 8. The cube root of 64 is 4, the cube root of 8 is 2, and finally I get x is equal to $4 - 2$, which is 2, which is correct. The formula works in a case where I sort of knew what the answer was ahead of time.

But, the formula works for others that aren't quite as nifty. That one all worked out; everything was a perfect square. Let me show you one where it's not quite as pleasant: $x^3 + 6x = 10$, m is 6, n is 10, and you go to the cubic formula. I'll read it once more, but this will be the last time. The cube root of $n/2$ plus the square root of $n^2/4$ plus $m^3/27$ minus the cube root of $-n/2$ plus the square root of $n^2/4$ plus $m^3/27$. You can sort of see at this point why they don't teach this in schools. It's hard enough to remember the quadratic formula, let alone this one. Anyway, if you stick in the m's and n's, you get a big blizzard of square roots and cube roots. When you simplify it, it comes down to be the cube root of 5 plus the square root of 33 minus the cube root of -5 plus the square root of 33. This can't be simplified further. That's it; the $\sqrt{33}$ is not something that reduces, so there's your solution by radicals. You could punch this in and get an approximate solution as a decimal; 1.300271 it comes out to be, but that's just an approximation. That's not what they were looking for in the 16th century. They were looking for the line above. They were looking for the solution by radicals.

A couple of questions remain. What if the cubic isn't depressed? What if you had a cubic that has the square term in it, like the generally $ax^3 + bx^2 + cx + d$ is 0. Cardano discovered how to approach this and what he did was he depressed it. If it wasn't depressed, make it so. His great discovery was how to do this, how to turn any cubic into a depressed cubic. He says you just make a substitution. You replace x by $y - b/3a$, where the a's and b's are the coefficients of your general cubic.

Let me do an example of this. Let me show you how this works. Let's suppose I have this cubic: $x^3 - 15x^2 + 81x$ is 165. That is not depressed. The a is 1, the coefficient of x^3. The b is -15. Cardano says let $x = y - b/3a$; that is $y - -15/3(1)$. Minus -15 is 15, 15/3 is 5, so he says let x be $y + 5$. When I go back to the cubic, I let x be $y + 5$, stick it into $x^3 - 15x^2 + 81x = 165$, and you'll get of course $(y + 5)^3 - 15(y + 5)^2 + 81(y + 5) = 165$. You multiply all of this out, collect the terms, and this turns into a depressed cubic. It always does; that's what Cardano showed. This always will have to happen. In fact, the depressed cubic it turns into is $y^3 + 6y$ is 10. That's the one we just did. I had x's in it, but now I've got y's in it. I know what the solution is by the formula y is equal to the cube root of 5 plus the square root of 33 minus the cube root of -5 plus the square root of 33, but that's y. What's $\underline{x}$? X is what

I want. Look up above; x is $y + 5$ so x, you just add 5 on, and there's your solution to the cubic, a solution by radicals. Even a non-depressed cubic can be solved by this process that Cardano discovered.

All of this appears in chapter 11 of *Ars Magna*, the chapter he called "Cubes and Roots Equal to Number"—$x^3 + mx = n$. But, remember in those days they had to avoid negatives, and so the example I just gave you that had a negative in it wouldn't have been legit in the 16th century. He had to do different chapters; each chapter would handle cases in order to avoid negatives. There were over a dozen different chapters to handle all the different possibilities for the cubic that we can do in a single expression because we're not nervous about negatives. The example I just showed you, $x^3 - 15x^2 + 81x = 165$, would not have been there. Rather you'd have to look in chapter 18, which tells you how to do cube and roots equal to square and number and sort of move this around a little bit and you'll see that this could be reduced to $x^3 + 81x$ cube and roots, bring the $-15x^2$ over to the other side, equals square and number. In chapter 18, you'd see how to do that one.

Cardano's *Ars Magna* continues to chapter 39 where we find Ferrari's solution of the quartic. Remember had shown how to solve a fourth-degree equation. Cardano introduces this with this nice statement. He said, "There is another rule, more noble than the preceding. It is Ludovico Ferrari's, who gave it to me at my request. Through it we have all the solutions for equations of the fourth power." *Ars Magna* solves the fourth degree. How does he do it? He reduces it to a third-degree. The third degree is reduced to a depressed cubic and away we go. Everything gets reduced downward.

It leaves one question hanging. What about the fifth degree? What about the quintic equation, $ax^5 + bx^4 + cx^3 + dx^2 + ex + f = 0$? How do you do the fifth degree? I know what I'd guess. I'll bet you'd reduce the fifth degree to a fourth degree, the fourth degree to a third degree, and solve it that way, but nobody could. People tried this in the 16th century; the fifth degree was a roadblock in the 17th and the 18th. Nobody could get anywhere, until the question was answered by Niels Abel in the 19th century, a Norwegian mathematician who showed that a solution by radicals of the general quintic equation, fifth-degree equation, or any higher degree equation is impossible; it can't be. There are no formulas for fifth-degree or higher. It's a very strange turn of events. You can solve the cubic and you can solve the quartic; after that, algebra fails.

I'll leave you with this thought that we sort of have verified the pessimism of Luca Pacioli. Remember way back in the beginning of this story, Pacioli said you couldn't do a cubic. He was wrong; you could do the cubic. His pessimism was unfounded. You can do the quartic, but you can't do the quintic or any higher. At that point, algebra fails us.

The Heroic Century
Lecture 12

Descartes' writing is regarded as being particularly opaque, particularly hard to follow. He himself seemed to take pride in this for some reason. … Descartes argued that only by making the reader struggle would the reader truly learn the material. It was his goal, his job, to make it hard to follow.

The 17th century is sometimes called the heroic century in mathematics because this period saw the development of many ideas that are now fundamental to our understanding of mathematics and many that are still being studied and researched.

Our first figure from this period is François Viète, who gave us a more modern algebraic notation in his book *The Analytic Art* (1591). As we've seen, earlier mathematicians had to express their equations in long, complicated sentences. Viète introduced the practice of allowing letters to stand for what had been termed roots, squares, and cubes.

Pierre de Fermat is perhaps the greatest mathematician from the heroic century.

Logarithms were another great innovation from this time, discovered and developed by John Napier and Henry Briggs. Together, these men created what are called the common, or Briggsian, logarithms. For example, the $\log_{10}$ of 100 is 2, because 100 is 10^2; the $\log_{10}$ of 1000 is 3 because 1000 is 10^3. In 1624, Briggs published *Arithmetica Logarithmica*, which was a table of logarithms that he had painstakingly calculated to 14-place accuracy.

Another significant achievement came from the well-known philosopher René Descartes. In an appendix to his *Discourse on Method* (1637), Descartes gave us the first published account of analytic geometry, that is, the idea of applying algebra to geometric figures in the plane. He also gave us modern algebraic notation, as we see in his expression of the depressed cubic.

Another Frenchman, Blaise Pascal, invented a mechanical calculating machine in 1642. Pascal was also one of the first people to transform probability into a mathematical science. One of his specific interests was the quadrature of the cycloid. The quadrature is the area under a curve, and a cycloid is a type of curve traced by a point on a moving wheel. Perhaps this mathematician is most well known for the development of Pascal's triangle, an array of numbers that is used in expanding binomials. For example, if we wanted to cube the binomial $a + b$, we could find the coefficients in row 3 of Pascal's triangle $(1, 3, 3, 1)$ and, thus, get the expansion: $a^3 + 3a^2b + 3ab^2 + b^3$.

Pierre de Fermat is perhaps the greatest mathematician from the heroic century. Fermat corresponded with Pascal on the development of probability theory, created his own version of analytic geometry, and foreshadowed both differential and integral calculus. His primary achievement was his work in number theory. The Fermat factorization scheme, for example, is simple but very clever. According to this, if we want to factor n, we let a be the largest whole number that is greater than or equal to $\sqrt{n}$. Then, we look at differences: $a^2 - n$, $(a + 1)^2 - n$, $(a + 2)^2 - n$, and so on. At some point, when we take the square of a number and subtract n, we will get a perfect square. We can then rearrange the problem into a difference of squares and factor it algebraically; the results will provide the factorization of n.

One other result of Fermat is motivated by the fact that sometimes two squares sum to a square; for example, $3^2 + 4^2 = 5^2$. According to Fermat, this is impossible with higher powers. In his notes, Fermat wrote that he had a proof of this assertion, but the margin in which he was writing was "too narrow to contain it." In fact, it wasn't until 1995 that this result, known as Fermat's last theorem, was finally proved—and the proof required hundreds of pages! ∎

Suggested Reading

Boyer, *A History of Analytic Geometry*.

Descartes, *The Geometry of René Descartes*.

Devlin, *The Unfinished Game*.

Hald, *A History of Probability and Statistics and Their Applications before 1750*.

Ore, *Number Theory and Its History*.

Weil, *Number Theory*.

Questions to Consider

1. Use Fermat's factorization scheme to factor 2,373,793 into the product of two smaller numbers. Start with the fact that $\sqrt{2{,}373{,}793}$ falls between 1540 and 1541. NOTE: Thanks to Fermat's insight, this is much easier than it looks.

2. Another of Fermat's number theoretic results is called "the little Fermat theorem." This says that if p is a prime and a is any whole number, then p divides evenly into $a^p - a$.

   (a) Check this numerically for the (easy) example of $p = 3$ and $a = 5$.

   (b) Now check it for the (considerably less easy) example of $p = 17$ and $a = 2$.

The Heroic Century
Lecture 12—Transcript

We've done battle with the cubic equation whose solution was the great achievement of the 16th century. Now it's time to move on to the next, to the 17th century, which is sometimes called the Heroic Century in mathematics. That's because the 17th century contributed so many ideas, which are now fundamental to our mathematical world, ideas that you encounter in school, ideas that are still being studied and researched to this day. It was the century when mathematics became modern. I want to introduce you to a cast of characters from at least the first half of the century in this lecture.

Perhaps first, though, we should just see what was going on elsewhere in the world early in the 17th century. There are a couple of pretty substantial achievements. In 1601, Shakespeare wrote *Hamlet*. In 1607, the British formed their first settlement in Jamestown in the New World. In 1609, Galileo turned his telescope on the heavens in the process confirming Copernicus's theory that the sun was the center of the solar system and discovering many other wonderful heavenly things as well. It was an exciting time, as it was in mathematics.

My first character from the Heroic Century is actually someone that lived most of his life prior to that. It was François Viète. Here's a dashing looking fellow; this is Viète. His great achievements were just at the end of the 16th, just before the 17th. These centuries don't divide perfectly in terms of ideas; I'm going to slosh him over into the 17th century. The debonair Viète did lots of things. One of them was he cracked the Spanish cipher. The Spanish army had a code, a cipher, in which they would encrypt important messages; the French couldn't break the code. They got a copy of one of these messages, they gave it to Viète, and with his great mathematical and logical ability he broke the cipher. He was one of the first codebreakers, therefore, but it's not for that that we're remembering him.

Rather, in his book *The Analytic Art* published in 1591, almost, but not quite in the 17th century, Viète gave us a more modern algebraic notation. Remember as we've seen all these people solving equations from Al-Khwārizmī to Cardano what they really were doing is working with words.

They had to express their equations in words and I've shared with you some of the very complicated verbiage that comes out of this when you're trying to solve an equation and you have to express it verbally. It's hard enough to do with all of our modern symbols. But, obviously mathematics was going to get stuck unless somebody came up with a more efficient way of expressing these things. Viète was the first to at least move us in that direction. In *The Analytic Art* you would find something like this; he would write D in R – D in E aequabitur A quad. This is an expression from Viète. It's not modern; it doesn't look exactly modern, but notice we have some letters playing the roles of what used to be called roots and squares and cubes. It's getting there. If you wanted to translate this particular expression into modern symbolism, what he's saying is D in R means $D \times R - D$ in E, which would mean $D \times E$, aequabitur, you could guess that that's equals, and A quad is A^2 as in quadratic. This is $D \times R - D \times e = A^2$. It wasn't perfectly modern, but algebraic notation was at least on its way. I'll show you in a few minutes the person that really modernized the notation, so that when we look at it when can read it, we can see what was being said algebraically, but that's coming a little later on.

Another great invention from this time and now we finally are properly in the 17th century was logarithms. You probably remember studying logarithms in school, one of the great, great discoveries of this century or any. Logarithms were discovered and developed by two people. One was John Napier, a nobleman from Scotland; there's his picture. One was Henry Briggs from England who was not a nobleman. In fact, I couldn't find a picture of Henry Briggs, try though I might. I asked for a picture of some English Henry from about this time and they gave me that one, Henry the VIII. Wait a minute, that's not Henry Briggs; that Henry wasn't particularly good at mathematics and he was a very unkind man. Henry Briggs was very good at mathematics and was very nice. He went and visited Napier and the two of them refined the idea of logarithms in a very impressive fashion.

They created what are called the common or "Briggsian" logarithms. These are the ones you remember from school. The $\log_{10}$, the common log or the Briggsian log, the base 10 log of 10 is 1. The $\log_{10}$ of 100 is 2, because 100 is 10^2. The $\log_{10}$ of 1000 is 3 because 1000 is 10^3. The logarithm is the exponent you must raise 10 to, to get the number. This is the idea of a common log or a Briggsian logarithm. Those are easy, but what about the $\log_{10}$ of 5? What number do you raise 10 to, to get 5? It's going to be some fractional number; it's not going to be very obvious what that is. It has to be calculated and it has to be calculated with an enormous amount of tedium, enormous amount of effort. Briggs took it upon himself to do this. He published in 1624 something called the *Arithmetica Logarithmica*, which was a table of logarithms, which he had calculated painstakingly over the years. It must've been one of the most grueling computations ever done because his tables of logarithms were done to 14-place accuracy, which seems like overkill to me, but finally he did this.

He presents the world with these log tables, and then people can use logarithms to expedite their calculations. If you remember the way this works, instead of having to multiply two numbers, you can just add their logarithms and use the tables. Adding is a lot easier than multiplying; it simplifies that. Instead of taking the cube root of a number, you can just take a third of its logarithm, and taking a third of a number is a lot easier than taking a cube root. This simplifies the kind of computational procedures radically. Logarithms were a fabulous timesaver. In fact, it was Laplace, the 18th-century mathematician, who wrote that logarithms "...by shortening the labors, doubled the life of the astronomer." Astronomers who had to spend so much time with their calculations could go so much faster with logarithms that they essentially doubled their career. Logarithms were a big achievement of the century.

Another big achievement came from this fellow René Descartes. He's very famous; we've hit a name everybody knows, Descartes. He was certainly a philosopher famous for his philosophical writings; probably chief among these was his *Discourse on the Method* of 1637. This is often viewed as the philosophical underpinnings of the scientific revolution that's on its way. Descartes is probably regarded by most people as a philosopher. True, but we want to also claim him because he was also a very substantial mathematician.

One of the things he did mathematically was attach an appendix to his *Discourse on the Method*, which was called *Geometry*. In this he gives us the first published account of what we now call analytic geometry. The idea that you will take geometry in the plane and superimpose a grid and apply algebra to the geometric figures, apply the geometry to the algebraic equations. You're going to connect algebra and geometry in what has to be the most glorious marriage in the history of mathematics. These two meet up, create analytic geometry, and Descartes tells us about it in his 1637 *Geometry*. He tells us, "…I shall not hesitate to introduce these arithmetical terms into geometry, for the sake of greater clearness." Indeed, where would mathematics be without our axes, without our functions, our curves, graphed within them? In fact, we call this now the Cartesian plane, the geometrical plane with the axes imposed, of course, named after Descartes.

He also gave us the modern algebraic notation. I mentioned that someone produced notation finally that looks modern; it was Descartes. Let me give you an excerpt. Here is a sample from Descartes's writing in which he has a formula. He's writing in French, you notice, not in Latin, which was kind of interesting. If you read what he's saying, he has an equation up there at the top or an expression that has a z^3 and a funny little symbol that looks like a fish and then $-pz + q$. He's talking about a rule, which Cardano had attributed to someone named Scipio Ferreus, and gives a root to that equation as follows and that's the big expression along the bottom. If you look at this, you can begin to see things. This looks kind of familiar. Let me examine this a little more carefully here. That little fish-like symbol up on the top line is how Descartes would write equals. On that front he wasn't exactly modern yet, but that is the equals sign. It took another century or perhaps a little less before the common equals sign became the standard.

Whenever he has z^3, little fish symbol $- pz + q$, what he's really saying is he's looking at $z^3 = -pz + q$. If I bring the minus pz over, what he's really looking at is $z^3 + pz = q$. That should look familiar; that's a depressed cubic. Instead of m and n as I was talking about in the previous lecture, we have p and q, and instead of x, I have z, but that's no big deal; that's the depressed cubic. That's why he introducing the name Cardan in his expression, which of course is Cardano, who got the rule from someone named Scipio Ferreus, and that's Scipio del Ferro. The rule itself is across the bottom. If you look at that expression on the bottom, that's the solution to the cubic. It takes a little bit of translation. For instance, both of those radicals have a radical and then a C period. What's that? That means cube root. He took the root with a C. for the cube. He didn't yet have the little 3 up there that we write, but that's not a big deal.

What else do we notice if we look at his expression? You see this thing, a plus sign with a line through it. A plus sign with a line through it—what's that? That's a minus. For some reason, Descartes hadn't used the minus, but he just lined out the plus. You see for instance between the two cube roots, there's a plus sign with a minus; if you remember the solution to the cubic, sure enough there's supposed to be a minus there.

One other thing that's a little strange is you see qq; you see that appearing twice in this expression. What's that? That's q squared. It's kind of interesting. In the 17th century, they didn't tend to write q^2. They would write qq; they'd write q twice. Once you got to q cubed, they didn't write qqq; they really wrote q with an exponent 3 and you'll see a p cubed at the very last expression up on the screen. For cubes and higher power, they use the exponents; for squares, they just copied the letter twice. I guess the reason was that you didn't save any space particularly in your printing to have qq as opposed to q^2. It still took two spaces, but q^5 only took two spaces, where $qqqqq$ would certainly start to fill up the line. In any case, when I look at Descartes, I see familiar looking mathematics. Finally, we have an algebra that we're familiar with.

Descartes's writing is regarded as being particularly opaque, particularly hard to follow. He himself seemed to take pride in this for some reason. He said, "I have omitted a number of things that might have made it clearer, but I did this intentionally and would not have done it otherwise." Here's a textbook author bragging that he's made his book hard to follow. Descartes argued that only by making the reader struggle would the reader truly learn the material. It was his goal, his job, to make it hard to follow. Modern textbook writers I would argue should not follow that advice. In fact, we're told that Newton later in the century tried to read Descartes. Newton's biographer wrote that Newton "...took Descartes' *Geometry* in hand, though he had been told it would be very difficult ..." Newton plunged in; he could get through one page and he got stuck. Then Newton said he'd go back to the beginning and this time he could get through two pages before he got stuck. Then he went back to the beginning and he got a little further before he got stuck. Then he went back to the beginning and eventually, says Newton, he got it. He mastered it, but it took all these starts—and this was Newton! If Newton had trouble with Descartes, look out; we're all in big difficulty here. Anyway, Descartes had given us analytic geometry and modern algebraic notation.

The next mathematician I want to talk about is another Frenchman, Blaise Pascal. Pascal was also known for something other than his mathematics. He was a prodigy who was great at all sorts of things—science, philosophy, theology—he was way ahead of his years even as a young person. At age 19, he invented a calculating machine, a mechanical calculating machine, now called the pascaline. Here's a picture of a recreation of this. Of course, our calculators are made out of silicon chips and wires and plastic. In those days, a calculating machine had to be made out of wood and metal, with cranks and gears. It was quite a complicated device, but Pascal's supposedly could add and subtract. You put the number in, turn some dials, and get the sum or difference. This was a great mechanical advance.

Pascal was known as a theologian. He wrote the *Pensées* in the 1650s, a book about his thoughts about theological matters. This is still being read. This is a work that has survived, even as Descartes's *Discourse* has survived. He was a physicist; Pascal worked on things like fluids, barometers, the hydraulic press. He was also a hypochondriac. Poor Pascal was always sick. He was always complaining about his health. He actually died young, so he might've had some legitimate reasons for those complaints, but you would expect he would've had an even more glorious career had his health been more robust.

But, it's in mathematics of course that we want to look at his contributions and there he worked in probability theory. He was one of the first people who started to turn probability into a mathematical science. I had mentioned Cardano had written the book on how to gamble, which was the first at least brush with probability theory, but Pascal's interests were a little more theoretical. He worked on the quadrature of the cycloid. This was a real important hot area in the 17th century. Quadrature, I had mentioned in an earlier lecture, means the area under a curve and the curve he's talking about here is the cycloid. If you don't know a cycloid, we're going to have to beat this. Let me tell you what this is.

Suppose you have a horizontal line and a circle upon it; think of a bicycle wheel going down the street. You take a look at one point on the wheel, on the circle. For instance, on a bicycle it would be the valve where you put the air in. Just keep your eye on that valve as the bicycle moves, so the valve will rise and fall and rise and fall and form these arches, as the bicycle goes down the street. There's a picture here showing the path traced by a point on the circle as the circle rolls along. That curve, those arches are called the cycloid arches, the cycloidal curve. This became a very important curve in the 17th century. People kept coming back to it, exploring it, investigating it, and finding strange and wonderful uses of it. Cycloids are quite important and it was Pascal who sort of elevated this to such an important position in mathematics. These were called mechanical curves, curves that could be created by something moving along.

He also gave us Pascal's Triangle. I guess in mathematics he's probably best known for that. Pascal's Triangle is an array of numbers; I have it up here. It's got 1 in the top, the next row is a 1 and a 1, and then you have a 1, 2, 1, and a 1, 3, 3, 1, and a 1, 4, 6, 4, 1, and so on as you move down through the triangle. You probably know the rule for this. To figure out what any number is, you add up the two numbers, the one above and to the left, the one above and to the right, and they generate the next number down. If I look at the bottom row there, there's a 15. Where did that come from? It came from the two numbers above, 5 + 10, so 5 + 10 is 15. That generates this triangle.

Is it just a pretty array of numbers? No, it has a use. It has a use in expanding binomials. If I took $a + b$, a binomial, two terms, and cubed it, so you multiply $(a + b)(a + b)$ and you get an answer and then you multiply once more by $a + b$, you get $a^3 + 3a^2b + 3ab^2 + b^3$. You can see what's happening there. The a's exponents are going down, a cubed, squared, a constant, the b's are going up, but the real trick is figuring out those coefficients and they're 1, 3, 3, 1 as you read across. That's just exactly row 3 of Pascal's triangle. What they're giving you are the coefficients that you would get when expanding by binomials. If I wanted $(a + b)^5$ and I didn't feel like multiplying that out five times, I don't have to. I just go to row five, I see the coefficients 1, 5, 10, 10, 5, 1, and I fill in the powers of a and b, and I get the expansion, $a^5 + 5a^4b + 10a^3b^2 + 10a^2b^3 + 5ab^4 + b^5$. That's it; it's much easier to do just by looking across that row. This is called Pascal's triangle. Truth to tell, Pascal wasn't the first person to discover this. The same array shows up in Viète's *Analytic Art* from 1591. Truth to tell, it actually shows up in ancient Chinese mathematics, so they were on to this well before Pascal, but his name got associated with it.

My last mathematician from the Heroic Century is the greatest of them, Pierre de Fermat. He makes up the third of the triumvirate of French mathematicians Descartes, Pascal, and Fermat. He was a very great mathematician indeed. His actual job however was to be a magistrate in Toulouse in France. This is a minor legal official. He was supposed to be working in the courts and apparently he didn't work very hard because he sure had a lot of time to do mathematics and he did. He did some great stuff.

One thing he did was he corresponded with Pascal on probability; remember I said Pascal had helped develop the theory of probability. He did so in conjunction with Fermat; they would write letters back and forth asking each other questions elaborating on their theories, and these letters formed the basis of probability theory. It's not just Pascal, but sort of you could say the Pascal and Fermat correspondence is what got probability going. That's pretty important.

He created an analytic geometry of his own before that of Descartes. Remember, Descartes published analytic geometry, but actually Fermat had done this long before, but he hadn't published it. Descartes gets his in print and we have the Cartesian plane. Had Fermat done his and published his earlier, we might've had the Fermatian plane or something. But, it was the same idea, that same very fruitful marriage of algebra and geometry. Fermat foreshadowed both differential and integral calculus. There's a pretty big achievement. Calculus was kind of almost ready to hatch, almost ready to appear, and Fermat has work where you can see him sort of getting real close to differential calculus, getting real close to integral calculus. He can find slopes, which is what differential calculus is about. He could find areas, which is what integral calculus is about. He didn't quite get there. He's not the creator of calculus, but he was real close.

His main achievement, that for which we remember him most clearly, is his work in number theory. We've seen number theory before back in the time of

Euclid. Euclid was a great number theorist. Fermat was a great, great number theorist. He did much with number theory, pushed the frontier greatly. I want to just show you one of his results and mention another from this rich mathematical field.

Here's the one I'm going to show you. It's called the Fermat factorization scheme. This is very clever. It's very simple, but it's very clever. Suppose I ask you to factor a whole number, because that's what number theory is about, into a product of two smaller pieces and I gave you 24. Can you write 24 as the product of two smaller pieces? Sure, that's easy: 4×6 or 8×3 if you want. There're lots of ways to do it. That's not very hard and it isn't. What if I give you this number and ask you to factor it: 1,940,249. Factor that into the product of two things. Hurry up—what would you do? It would be very complicated. I don't know what the factors are; it's not obvious. It would be nice to have a technique, a scheme that would efficiently and quickly give me the factors. That's what Fermat describes. It's very slick. Let me show you.

He says if you want to factor a number N, capital N, here's what you do. You let a be the largest whole number that's greater than or equal to the $\sqrt{N}$. Then you start looking at these differences. You look at $a^2 - N$, you look at $(a + 1)^2 - N$, that's the next number up, squared minus N, you look at the next number up $(a + 2)^2 - N$, and you keep going. You look at $a^2 - N$ and then all the bigger numbers minus N. Fermat says suppose at some point when you take the square of a number minus N, you get a square; suppose the difference at some point is a perfect square. In other words, let's say you climb all the way up to c, so you have $c^2 - N$ and it turns out to be b^2. If you get to that stage, then you just do a little rearranging, N would be $c^2 - b^2$, but $c^2 - b^2$ is a difference of squares. You can factor it algebraically into $(c - b)(c + b)$, and then the numbers that will go in there will provide the factorization of N.

Let me illustrate it by factoring my monster number 1,940,249; let's see this get factored. According to Fermat's scheme, what you do first is take the square root of 1,940,249 and that's 1392 and change. What you do is you go up to the next bigger number, so that's 1393. Then, you take 1393^2 minus your number, 1,940,249. You go up to the next one and you get 1394^2 minus your number, 1395^2 minus your number, and keep going until you get a perfect square as the result. In my case, that first difference turned out to be 200; 1393^2 minus my number is 200, and that's not a perfect square. The next one turned out to be 2987—no, that's not a perfect square. But, the next one, 1395^2 minus my number, turned out to be 5776, which lo and behold is 76^2. That's all the further I've got to go because now I see that $1395^2 - 1{,}940{,}249$ is 76^2. Move this around and you'll get that 1,940,249 is therefore $1395^2 - 76^2$. That's a difference of squares; that factors into $(1395 - 76)(1395 + 76)$. You add and subtract here, and you see that it's 1319×1471. That's how to factor 1,940,249 with Fermat's little trick, and notice it only took three tries to do it. Wow, you can factor that on your third try. That's nice.

The other result of Fermat is motivated by the fact that sometimes two squares add up to be a square as we know, $3^2 + 4^2$ is 5^2; 5^2 is 25, and you can sort of break it down into two squares. Fermat said it is impossible to do this with higher powers. He said, "… it is impossible to divide a cube into two cubes, or a fourth power into two fourth powers, or generally any power beyond the square into two like powers. I have found a remarkable demonstration, but this margin is too narrow to contain it." He was writing this in the margin. He said that you can't split a cube up into two cubes, nor a fourth power into two fourths, nor any higher power. He had a proof, but he couldn't quite fit it in. What he's saying in our notation is if n is bigger than 2, there are no whole numbers a, b, and c so that $a^n + b^n$ is c^n. For squares you can do it, but not for any higher N.

This result has come to be known as "Fermat's Last Theorem." It was a tantalizing problem for mathematicians, in part because he claimed he had a solution—he just couldn't quite get it in the margin, maybe somebody could find it. Nobody could in the 17th century, nor the 18th, nor the 19th, and indeed this wasn't solved until the year 1995 by Andrew Wiles with an assist from Richard Taylor. Fermat was right, but the proof that they found was hundreds of pages long. It would not have fit in the margin.

I'll end this lecture with a quotation from Fermat where he was looking back over his work and I think quite proud of what he had done. He wrote this: "Perhaps posterity will be grateful to me for having shown that the ancients did not know everything." That kind of captures the excitement of the era, the excitement of the Heroic Century.

The Legacy of Newton
Lecture 13

It's been said that Newton's *Principia* is the greatest science book ever written, and that might well be true. If it has any rival, it would be Darwin's *Origin of Species* from centuries later.

Isaac Newton was born in 1642 to a widowed mother in a small town in Lincolnshire. As a child, he was said to have been sober, silent, and very smart. In 1661, he entered Trinity College, where he was mentored by Isaac Barrow, the Lucasian Professor of Mathematics. Barrow directed Newton's readings into Descartes and other mathematical thinkers.

The years 1665–1667 were the *anni mirabilis*—miraculous years—for Newton, the time when he pushed the frontiers of mathematics and science. In this period, he discovered the generalized binomial theorem, differential and integral calculus, and the laws of motion—all while a student at Trinity College. When the college had to shut down because of an outbreak of the plague, Newton went home, where he developed the theory of universal gravitation. According to contemporary descriptions of Newton, he became so caught up in his work that he often neglected to eat or sleep and never engaged in any outside recreation or pastime. Nor was Newton in any hurry to share his discoveries with others. He believed that publication would bring with it an increase in social activities, which he desired to avoid.

In 1669, he was appointed to succeed Barrow as Lucasian Professor of Mathematics. At about the same time, he invented the reflecting telescope, which he sent to the Royal Society in London in 1671. As a result, Newton was made a member. The next year, he submitted some of his papers on optics to the Royal Society but received criticism from the scientist Robert Hooke. For much of the remainder of the 1670s and 1680s, Newton abandoned mathematics and science to devote his time to alchemy and theology.

In the later 1680s, with the urging and financial backing of Edmund Halley, Newton finally began to publish some of his work in mechanics. In 1687, Newton's great work the *Principia Mathematica* appeared. Here, he described the laws of motion and gravity in a very Euclidean fashion, putting forth definitions and axioms, then deducing propositions. The *Principia* launched Newton into fame, whether he wanted it or not.

In 1689, Newton was elected to Parliament, but his government career was brief and not particularly memorable. In the years 1692–1694, he seems to have suffered a period of mental derangement, perhaps brought on by mercury poisoning in connection with his work on alchemy. Fortunately, he recovered and, in 1696, became warden of the mint. He apparently was quite successful in this position, overseeing the recoinage in Britain and prosecuting counterfeiters with great zeal.

In 1687, Newton's great work the *Principia Mathematica* appeared. Here, he described the laws of motion and gravity in a very Euclidean fashion, putting forth definitions and axioms, then deducing propositions.

In 1703, Newton was made president of the Royal Society, a position he held for the rest of his life. In 1704, he published his *Opticks*, a work on light and color. In 1705, he was knighted by Queen Anne. By this point, Newton was quite famous and wealthy, but he remained crotchety and contentious throughout his life. His scholarly disputes included those with Robert Hooke; John Flamsteed, the royal astronomer; and Gottfried Wilhelm Leibniz, the German mathematician and philosopher who invented calculus independently of Newton and at around the same time.

Newton died in 1727 and was buried in Westminster Abbey. Voltaire reported that the British "buried him as though he had been a king …," and Wordsworth later described the statue of him that stands in Trinity College chapel: "Of Newton with his prism and silent face, / The marble index of a mind for ever / Voyaging through strange seas of Thought, alone." ■

Suggested Reading

Fauvel, et al., *Let Newton Be!*

Gjertsen, *The Newton Handbook.*

Hall, *Philosophers at War.*

Newton, *The Correspondence of Isaac Newton.*

Westfall, *The Life of Isaac Newton.*

———, *Never at Rest: A Biography of Isaac Newton.*

Questions to Consider

1. To get a sense of the era, read about mathematician Isaac Barrow (Newton's mentor), physicist Robert Hooke (Newton's nemesis), and counterfeiter William Chaloner (Newton's prisoner).

2. Newton lived during a turbulent time in the history of England. He was born in 1642, at the start of the English Civil War, which culminated, in 1649, with the execution of King Charles I; he began his Cambridge career the year after the monarchy was restored under King Charles II; and by mid-career, Newton saw King James II flee the country for reasons of religion. Read up on these times, and imagine how a person like Newton would have reacted to all the turmoil.

The Legacy of Newton
Lecture 13—Transcript

In my previous lecture, I talked about mathematics from the 17th century, the Heroic century. We met people like Descartes and Fermat. In this lecture, we're going to meet the Heroic century's greatest hero, Isaac Newton. Let me begin with a picture. This is Newton. This portrait is sometimes called *Newton in His Own Hair* because he wasn't wearing a wig. This is actually what he looked like.

I want to trace his life and tell you about his character in this lecture. In later lectures, we'll look at some of his actual mathematical work. Newton was born in 1642 in Woolsthorpe, England, a small town in Lincolnshire, north of Cambridge, far north of London. His life started very precariously. His father died before Newton was born so he had no father to help raise him. More seriously perhaps, he was dangerously premature. He came way too soon and it was said that the infant Newton was so tiny he could be put into a quart pot. In the 17th century, that was almost certainly a death sentence. However, little Isaac was a tough kid. He survived and lived a very long life.

Those who knew him remembered that, as a child, he was a sober, silent, and thinking lad—not particularly gregarious, but very smart. He didn't seem to be cut out for business. He didn't seem to be cut out for farming. What do you do with such a person? Ah, you send them to university. In 1661, Isaac Newton enters Trinity College Cambridge. Today, Cambridge is one of the great institutions in the world certainly. What he found in 1661, however, was not quite that. In that time, Cambridge was still teaching the medieval curriculum. He would still be worried about translating Aristotle from Greek and working on Latin sentences and things of that sort. You weren't paying attention to the modern inventions of science and the modern discoveries of mathematics that were taking place on the continent; that just hadn't penetrated yet.

Fortunately, it was also the case that the people at Cambridge didn't much care what the students studied, what they did. They could pretty much go off on their own. This is what Newton decided to do. He had a mentor, a fellow named Isaac Barrow who was the Lucasian Professor of Mathematics at Cambridge. Barrow wasn't Newton's teacher in any formal sense, but he would direct Newton to read that, to read Descartes, and to look at this particular mathematical result. Pretty soon Newton working alone zooms to the forefront.

The years 1665–1667 were called his *Anni mirabilis*, his miraculous years, when he's working at a fever pitch pursuing his mathematics and science, pushing the frontiers beyond what anybody had ever done before. Just a few examples, what did he do during these miraculous years? He discovered the generalized binomial theorem. This becomes his entree into higher mathematics and I'm going to talk about this in the next lecture. What else? He discovers something he called fluxions. We know it by the name of differential calculus—not bad for an unknown student. He discovers inverse fluxions, integral calculus. He discovers properties of the laws of motion; he works on optics; he does all of these things that are just phenomenal while working alone at this fever pitch in his rooms at Trinity.

There're descriptions from the time of Isaac sitting at his desk writing away in the evening and the candles are high. Someone comes back and sees him in the morning. He hasn't moved. He's still writing, but the candles have burned low. He forgot to sleep; he forgot to eat. It was said that his cat grew fat eating Isaac's untouched meals. Rarely, maybe never, did somebody devote themselves so passionately to learning and research. If you don't believe me on this passion, let me show you a page here from Newton's notebook from the time. He was studying here the properties of light and color and vision. At that time these were all sort of one subject. Nowadays we sort of take light and color and put them into physics and vision shows up in biology, but he was studying these as a unit. He wanted to see how the eye worked vis-à-vis images.

What he tells us in this notebook passage is, "I took a bodkin"—a bodkin is a sharp little stick, think of a nail file, a little knife. "I took a bodkin," he said, "and put it betwixt my eye and the bone as near to the backside of my eye as I could." To see how the eye works, he's taken this knife and stuck it behind his eye. In that page I'm showing you there's a picture of his hand with a knife in the eye and he tells us that he made the curvature of his eye *a, b, c, d, e, f* and he has dutifully labeled this *a, b, c, d, e, f*, and you see the eyeball distended. Then he said when he pressed this bodkin he saw these white, dark, and colored circles r, s, and t, and if you look at the picture out there are these little circles. He says when he rubbed his eye with the bodkin, the circles would grow bright, but if he left the bodkin still and didn't move them, they'd kind of fade away. How does the average student study light? They stick a knife behind his eye. No, do not do this at home. This is dangerous, but this is how you do it if you're passionate about learning. This is what Newton did. I can't prove it, but my hunch is about this time his cat grew queasy.

In 1665–67, the miraculous years, the plague struck Cambridge and Newton had to go home to Woolsthorpe. The university was closed for health reasons. While there he later tells us he was outside and he saw an apple fall from a tree. He said there's some force pulling that apple to earth; mighten that force extend out into space? He wrote later, "… I began to think of gravity extending to the orb of the moon … and I deduced that the forces which keep the planets in their orbs must [be] reciprocally as the squares of their distances from the centers about which they revolve; and thereby compared the force required to keep the moon in her orb with the force of gravity at the surface of the earth, and found them to agree pretty nearly." Here's Newton hatching the idea of universal gravitation. He tells us that, "… in those days I was in the prime of my age for invention and minded mathematicks and philosophy more than at any time since." Did he ever! What he was doing here was charting the course of modern science and doing it singlehandedly as an isolated student.

His biographer Richard Westfall writes this description of Newton during these incredible years. He said Newton's was "… a virtuoso performance that would have left the mathematicians of Europe breathless in admiration. As it happened, only one mathematician in Europe, Isaac Barrow, even knew that Newton existed." It's kind of a chilling thought that here was the greatest scientist alive, working completely unknown. Only Barrow, his mentor, even knew he was there.

As far as his personality, one of his contemporaries remembered, "I never knew him to take any recreation or pastime, either in riding out to take the air, walking, bowling, or any other exercise whatever, thinking all hours lost that were not spent in his studies." No time for exercise; you've got to work, and you've got to think. It was said that, "He very rarely went to dine in the hall … and then, if he had not been minded, would go very carelessly, with shoes down at the heels, stockings untied, … and his head scarcely combed." Here's the absent-minded professor finally being told you better go eat something and he forgot to tie his shoes or comb his hair. This was Newton. It's too bad because the hall mentioned in that passage is this. This is the dining hall at Trinity. It's a beautiful building; it sits on Trinity Great Court, which is probably the most glorious academic setting in the world. Newton should've gone through there, gone to the hall more often, and enjoyed the scenery.

We're told also that he was humorless; he was very sober. A friend, a colleague, who had known him for decades was asked once did you ever see Isaac Newton laugh? The friend thought about it and said once, one time, Isaac Newton laughed. The obvious question is what made him laugh? What made Newton laugh for this one and only time? The friend reported that someone asked him if there were any value to be had from reading Euclid's *Elements*. That got a chuckle out of Isaac—otherwise, sober all the time.

Newton was in no hurry to share his great discoveries with others. He's writing all these incredible things; he's just putting them in his desk, not publishing, not rushing them out into the academic world. He once said, "I see not what there is desirable in public esteem, were I able to acquire and maintain it… It would perhaps increase my acquaintance, the thing which I chiefly study to decline." He didn't want to publish; he didn't want to share this because then it would've made acquaintances and that would've got him all embroiled in social activities, all of which would deprive him of the time for his precious studies.

In 1669, Newton was appointed Lucasian Professor of Mathematics at Cambridge. This is a major upset. Remember, he's not published anything. He was unknown and now he gets one of the plumb jobs in England, the Lucasian Professor. How did this happen? It turned out that the Lucasian Professor previously was, as I said, Isaac Barrow. Barrow was off the court, off to London to take a promotion, so he was leaving the chair. But, he knew Newton and he knew how good Newton was and recommended that this unknown Newton fill the Lucasian chair. Barrow said Newton was "… a fellow of our college, and very young, but of an extraordinary genius and proficiency in these things." On Barrow's recommendation, Newton became Lucasian Professor. As such, he had very few responsibilities, but one of them was he had to give a lecture every once in awhile, a public lecture. Apparently, he was really bad at this. A contemporary remembers, "… so few went to hear him, and fewer yet understood him, that oftimes he did in a manner, for want of hearers, read to the walls." Newton was just speaking to an empty room. That's kind of amazing. Today, if you knew Isaac Newton was speaking on campus, you'd go just to hear him. Then when they had that opportunity, nobody went.

Newton invented about this time the reflecting telescope. Here's yet another of his many, many achievements. If you know telescopes, you know there are two types. There's the refractor, which is a tube with a lens at the far end, like a spyglass, and then there's the reflector, which is open at the far end, but has a mirror at the base that magnifies the objects. The latter kind, the reflecting telescope, is the one that all the big telescopes are, the Hubble and other telescopes around the world. It's much more efficient to put a heavy mirror at the base. This is Newton's invention and here's a drawing of it that he did. He was actually very good at making these things. He was very good with his hands and he ground the lens just right and made this neat little reflecting telescope. He then sent it down to the Royal Society in London in 1671. The Royal Society was the great scientific body in England and this is where all the hotshots were. Here comes the reflecting telescope from Newton; they loved it. They thought this was an incredible invention and they made Newton a member of the Royal Society. For the first time, he's sort of getting out; he's sort of sharing some of his discoveries. It looks like the beginning of a glorious career.

But, wait a minute. The next year in 1672, Newton sent down some of his optical papers to the Royal Society in which he's describing his theories of light and color. At this point, he gets criticism. Robert Hooke was at the Royal Society then; if you know physics, you might've heard of Hooke's Law, that's Hooke. Hooke objected to Newton's optical theories. He had issues and he criticized it. What's supposed to happen in science is you put forth the theory and somebody might critique you. You have a rebuttal and they have a rebuttal, and you try to refine your differences and build a theory that everyone can embrace. The ball was in Newton's court and he was supposed to respond. He doesn't. He says more or less, you don't like my paper, Mr. Hooke, I'm not sending you anything else. He withdraws from this. He once wrote, "There is nothing which I desire to avoid more than contention, nor any kind of contention more than one in print." He was not going to get into a battle with Hooke or anybody over his science. Newton was confident that he was right and he wasn't going to argue with these lowly people at the Royal Society.

What happens is there is a long period where he more or less abandons mathematics and science and devotes his time to alchemy and theology. For much of the 1670s and '80s he's working in a completely different arena. Alchemy as you know is this sort of proto-chemistry. It was this attempt to turn base materials into gold. Newton built a little laboratory to work on this. He built little furnaces. He got chemicals and he was trying to make gold out of other things. He wrote a million words on alchemy. It said he knew more about alchemy than anybody ever. None of this has survived; none of this is of any particular interest to anyone. I think we would regard it a waste of Newton's time.

Likewise he studied theology. He studied the Bible. He read it carefully. He was looking for kind of clues in one book to help him understand another. He would do things like calculate the size of the arc of the covenant using his mathematics and he was trying to discover prophecies of the future. He wrote a book called *Observations On the Prophesies of Daniel*. Like his alchemy, this is not read anymore; nobody much cares. Unlike Pascal who's *Pensées* is still read in theology, Newton's theology is just sort of drivel.

We can look at this and say why did you spend all this time on these tangential issues when you could've been advancing math and science in a way that only Newton did? But, that's the story. Finally in 1684, Edmund Halley was up in Cambridge. He was talking to Newton, and he begins to get a sense of what it was that Newton had in that desk, some of these great scientific ideas. Halley says to Newton, more or less, you've got to publish this. You simply can't sit on this any longer; this is earth-shaking stuff. Come on, Newton, you've got to publish.

At Halley's urging and indeed with some of Halley's funding, Newton publishes his mechanics. Finally the world is going to see Newton in action. That comes in 1687 in his great work the *Principia Mathematica*. Here's the title page, the Mathematical Principles of Natural Philosophy, 1687. This is Newton's mechanics. This is where he puts his laws of motion and the laws of gravity; he describes the system of the world. He does this in a very Euclidian fashion. He begins with his definitions, his axioms, and then he starts deducing propositions. It certainly looks familiar to a reader of Euclid whom he so much admired. But, he doesn't put in here his calculus, his fluxions. That's not part of the book. He's writing a very important physics tome, but not a very important mathematical one.

It's been said that Newton's *Principia* is the greatest science book ever written and that might well be true. If it has any rival, it would be Darwin's *Origin of Species* from centuries later. It's interesting that Darwin also was an alum of Cambridge University. There must be something in the water up there. In any case, with the publication of the *Principia*, Newton is now famous whether he wanted to be or not. This put him on the front page and his life changes. It changes in some very strange and unexpected ways.

In 1689, he is elected to Parliament. This seems such a surprise to us because for us politicians are very gregarious—they shake hands, they kiss babies. Newton of course did not do anything like that. What's going on here? How did he get into Parliament? The answer is that in 1688, the year before, Britain had experienced the glorious revolution. At that time, the Catholic King James II had been driven from the throne and the Protestants William and Mary had taken over. Newton was a dedicated anti-Catholic. He was glad to see James go and he wanted to support the new monarchs William and Mary, and thus stood for election to Parliament to represent Cambridge and won. He went down to London and served a term in Parliament. His government career was apparently not very good. He didn't do anything particularly notable, but he sort of got out. Finally Isaac is getting out into society and in fact he sort of liked it. He met John Locke, the great philosopher. He met Samuel Peeps, the diarist, and he sort of finally was getting away from this isolated life at Cambridge.

In 1692–94, he suffered his period of mental derangement. He suddenly cannot get out of bed. He doesn't answer mail. This most active, energetic of scholars seems to be suffering from lethargy. What's this? Nobody's certain, but there's a theory about what was going on with Isaac at this time and that is based on his alchemy work. When he was doing his alchemy, in order to tell whether his chemicals were pure, he would taste them. He gets a new shipment of mercury in and he wants to see if the mercury is any good, so he eats it. This is not good. This can affect your brain. It's possible he was poisoning himself with his chemicals and thus experienced this period of mental derangement. If you think it's strange that he would eat chemicals, remember that this is the same guy that put the bodkin behind his eyeball, so eating chemicals was nothing for him. Fortunately, he gets over his period of mental derangement.

In 1696 in another dramatic change, he becomes warden of the mint. He becomes a government official. The mint of course deals with the money supply in Britain. The mint was in the Tower of London. Newton moved to London and he would go to work in the Tower overseeing what was called the great recoinage, as they were minting new coins for Britain, working with the people in the Tower, working with the bankers in the city of London, working with the government running the mint, and apparently he was really good at this. He was a very good warden of the mint, very efficient, the coins got minted, and he was seen as a great guy. Also, while there, he was responsible for taking care of counterfeiters. As warden of the mint you were responsible for prosecuting counterfeiters and this he did with great zeal. There was a famous case; in 1699, a very well-known debonair counterfeiter named William Chaloner ran afoul of Newton. Chaloner was caught, Newton oversaw the prosecution, which led to a conviction for counterfeiting, and Chaloner was hanged. Newton was quite happy with this outcome. You didn't want to cross him at the mint.

In 1703, Newton is made president of the Royal Society. This is kind of interesting. The Royal Society where he had run into Robert Hooke so many years before, now he was running the show. He remained in the Royal Society as president for the rest of his life. In 1704, he publishes his *Opticks*, his other great work on light and color. In 1705, Newton is knighted by Queen Anne. She comes up to Cambridge, knights him at his old alma mater, and he's now Sir Isaac. He's living in London running the mint, running the Royal Society, very famous, and actually he gets very rich along the way, leading a life that he would never have anticipated as that solitary, young scholar decades before.

He remained crotchety and contentious throughout his life and he got into all sorts of scholarly disputes with people. I've already mentioned he got into a battle with Robert Hooke over the theory of optics. Let me mention two others. He went head to head with John Flamsteed. Flamsteed was the Astronomer Royal. He worked at Greenwich Observatory and he and Newton locked horns over astronomical data. It was very bitter and very unpleasant. Most famously, Newton went to war with Gottfried Wilhelm Leibniz, the great German mathematician and philosopher, who invented calculus on his own. There was in fact a calculus war that erupted between Newton and his followers and Leibniz and his. We'll talk about this in a later lecture.

But, as to his calculus, his fluxions, the great mathematical discovery of Newton's life, he does not publish that until very late. In fact, the complete treatment of Newton's fluxions doesn't appear in print until 65 years after Newton wrote it and nearly a decade after he died. It was actually a posthumous publication. Boy, it would seem to me if I had discovered calculus, fluxions, I'd have published it right away. It actually didn't come out until Newton was gone. This is what it looked like when it finally hit the streets. It's called *A Treatise of the Method of Fluxions and Infinite Series with Applications to the Geometry of Curve Lines by Sir Isaac Newton, Kt.* Of course, by then, the thunder had been stolen by others who had been writing about calculus for decades. But, I'll tell more about that story when we get to the calculus wars.

Newton lived until 1727, a very long life. When he died, he was buried in Westminster Abbey. Here's a picture of the west front of the Abbey. This of course is the great shrine of England. This is where the British bury their heroes, their kings and queens, their great statesmen, great soldiers, great writers, and Isaac Newton. If you enter through the west gate and look down the knave, right across the front you'll see something called the choir screen. It sort of bisects the knave. On the left side is a huge monument and it's the monument to Newton—here's a picture—and below that is Newton's grave.

You'll see there an image of Newton lounging with cherubs around him, the world above, and he's reading his *Principia*. He has been enshrined in the greatest spot of all, Westminster Abbey.

Voltaire was in London for Newton's funeral and he reported Newton's "... countrymen honored him in his lifetime and buried him as though he had been a king..." Here's a mathematician buried like a king, but it was well deserved. On the tomb is this inscription. It says: "Mortals, rejoice at so great an ornament for the human race." The British were paying homage to their great son Isaac Newton. People wrote poems about him; people carved statues of him. He was a figure of just heroic proportions at the time.

They were still honoring him in the 20th century on their money. The British one-pound note, when they still had a one-pound note, looked like this. I should just point out nowadays the British have a one-pound coin, and so this is old currency, but when they had the one pound note, on the one side was the queen, and on the other side was Isaac Newton. There he is in his wig. You see his reflecting telescope; you see his volume of the *Principia*. He gets on the one. I remember when I first saw this I was very impressed. The one was the most common currency in circulation. You could put your greatest hero on the one, so the British could've put Shakespeare on the one, but they didn't. They could've put Wellington, no—the Beatles, no. They put Isaac Newton, their greatest hero, on the one-pound note. It didn't hurt that he had been warden of the mint. I think he had the inside track there, but in any event that was certainly a testimonial to Newton.

Let me end this with a picture of a statue of Newton. This is in the Chapel at Trinity College Cambridge. He's holding his prism and gazing off in contemplation. This statue was there long ago when a young student named Wordsworth, William Wordsworth, attended Cambridge. Wordsworth went to St. John's College, which is right next door to Trinity College. From his rooms, Wordsworth could see the Chapel at Trinity. Wordsworth writes this poem. He says:

> And from my pillow,
> Looking forth by light
> Of moon or favouring stars,

I could behold the antechapel
where the statue stood
Of Newton with his prism and silent face,
The marble index of a mind for ever
Voyaging through strange seas of Thought, alone.

Wordsworth has captured the solitary scholar, the great hero, Isaac Newton.

Newton's Infinite Series
Lecture 14

> Newton's insight in 1664 was that you could use this expansion not just for whole numbers r but for other kinds of exponents. What he said is, r could be integral or (so to speak) fractional, positive or negative. … Nobody had ever thought of that.

The exploration of infinite series was at the cutting edge of research in 17th-century mathematics, and Newton was at the forefront of this study. In this context, "series" means sum, and an infinite series is the sum of infinitely many terms.

We begin with a finite series. We could use the FOIL method to solve the binomial $(1 + x)^2$, but for a problem like $(1 + x)^{17}$, we need something called the binomial expansion. Let's consider $(1 + x)^r$. The expansion starts with 1. When we multiply $1 + x$ by itself r times, all the 1's will multiply. We then get $r \times x$. For the third term, we get $r(r - 1)/2 \times 1 \times x^2$. The x^3 term has as its coefficient $r(r - 1)(r - 2)/3 \times 2 \times 1$, and the pattern continues. We are always marching down from r: $r - 1$, $r - 2$, and so on.

Newton realized that the binomial expansion could be used even if the exponent was not an integer.

Newton realized that the binomial expansion could be used even if the exponent was not an integer. As we go through the math, keep in mind that we can use the simplified notation 4! for an expression like $4 \times 3 \times 2 \times 1$. Also note that a negative exponent, such as x^{-n}, essentially means $1/x^n$. Likewise, a fractional exponent, such as $x^{p/q}$, means $\left(\sqrt[q]{x}\right)^p$.

Suppose we want to expand $(\sqrt[3]{1} + x)$ as a series. We first turn the cube root into a fractional exponent: $(1 + x)^{1/3}$. When we use the binomial expansion, the role of r will be played by 1/3, as follows: $(1 + rx)$ will become $1/3(x)$. Next comes $1/3(1/3 - 1)/2!(x^2)$, followed by $1/3(1/3 - 1)(1/3 - 2)/3!(x^3)$. Simplifying, we get $1 + 1/3(x) + 1/3(-2/3)/2 \times 1(x^2)$, because $1/3 - 1$ is $-2/3$. The next term will be $1/3 \times (-2/3)(-5/3)/3 \times 2 \times 1 (x^3)$, because $1/3 - 2$ is

–5/3, and so on. Cleaning up the fractions, we get $1 + 1/3(x) - 1/9(x^2) + 5/81(x^3) - 10/243(x^4)\ldots$. This is an infinite series.

Newton thought this tool was useful for finding roots. Suppose we were asked to find $\sqrt[3]{140}$. When we use Newton's generalized binomial expansion, we must be sure that the value of x we use is smaller than 1 so that the terms zero in on an answer. To find that value, we look for a perfect cube that is close to 140, such as 125, which is 5^3. We rewrite the problem as $\sqrt[3]{125} + 15$, then factor out 125: $\sqrt[3]{125}(1 + 15/125)$. Next, instead of $\sqrt[3]{125}$ multiplied by the parenthetical, we rewrite the expression as $\sqrt[3]{125}(\sqrt[3]{1+3/25})$. We break these apart into two cube roots, and what we're left with is $\sqrt[3]{1+3/25}$. We can now use the binomial expansion because x will be 3/25. If we put that value into the series, it will converge quickly to an answer. Stopping after the first four terms, we get: $1 + 1/3$ of x will be a 1/3 of 3/25 minus 1/9 of x^2 will be minus 1/9 of $(3/25)^2$ plus 5/81 of $(3/25)^3$. Collecting terms, we are left with 9736/9375, or 5.19253, as our approximation of $\sqrt[3]{140}$.

Newton also did the infinite series expansion of the sine, a well-known trigonometric ratio. His sine series is: $\sin x = x - x^3/3! + x^5/5! - x^7/7!\ldots$, which has become an important result in mathematics down to the present day. ∎

Suggested Reading

Boyer, *The Concepts of the Calculus.*

Dunham, *The Calculus Gallery.*

Edwards, *The Historical Development of the Calculus.*

Gjertsen, *The Newton Handbook.*

Grattan-Guinness, *From the Calculus to Set Theory.*

Westfall, Richard. *Never at Rest.*

Whiteside, ed., *The Mathematical Works of Isaac Newton.*

1. Use the first four terms in Newton's generalized binomial expansion to approximate $\sqrt[3]{72}$. Recall the trick: $72 = 64 + 8 = 64\left(1 + \frac{8}{64}\right) = 64\left(1 + \frac{1}{8}\right)$. Your approximation should be quite accurate.

2. Convert 30° to radian measure, then use the first three terms of Newton's series for the sine (i.e., $\sin x \approx x - \frac{x^3}{3!} + \frac{x^5}{5!}$) to approximate the sine of this angle. NOTE: As you might recall from trigonometry, the exact value is $\sin(30°) = 1/2$.

Newton's Infinite Series
Lecture 14—Transcript

Having met Isaac Newton, we should now take a look at some of his mathematics. I'd like to devote two lectures to Newton's math and this first one will be about his work with infinite series.

Infinite series was a hot topic in 17th-century mathematics. It was at the cutting edge of research and Newton was at the forefront of this study. I should say what is an infinite series. Series means sum, and an infinite series is the sum of infinitely many terms. The question before the mathematicians at that time was what happened to the series, did it add up to something useful, what sort of interesting properties did such series have? I want to show you what Newton did with this.

In the interest of starting slow let's start with a finite series and work our way up. I want to begin with a binomial. A binomial means two terms. A binomial is something like $x + y$, $a + b$, something of that sort. Let's take a look at ($1 + x$), that binomial, times itself, so $(1 + x)(1 + x)$. We know how to do this from high school algebra; they call it the FOIL method. You multiply the first terms, you get 1×1 is 1, $x \times 1$, the inner term, the outer term, and the last term, and then rather quickly it comes out to be $1 + x + x + x^2$. If you were to add those two x's in the middle, you get $1 + 2x + x^2$. What we're saying is that $(1 + x)^2$ is $1 + 2x + x^2$. There's nothing hard about that. But, what if we had to figure out $(1 + x)^{17}$ power? That's considerably more challenging. There are a couple of options. The worst one is to multiply $(1 + x)(1 + x)$ and get an answer and then multiply a third time by $1 + x$ and get an answer, and a fourth, and a fifth, and all the way up to 17 multiplications. This would be a very dull way to spend an afternoon. You could do it, but no, we don't want to go down that road. Is there a shortcut?

We saw one a few lectures ago when we talked about Pascal's triangle. Remember that was that array of numbers 1, 1 1, 1 2 1, 1 3 3 1, where each number in a given row is formed by adding the two numbers above it—one to the left, one to the right. You generate these rows and these numbers across the rows are exactly the coefficients you get when you expand something like $(1 + x)^{17}$. If I wanted to do that with Pascal's triangle, then I'd just go down to the 17th row of the triangle and I could read off the coefficients. It would be great except there's a problem there too because the 17th row requires me to have the 16th row. The 16th depends on the 15th, and so I would actually have to build from the start all the way down to the 17th row. That wouldn't be much fun either. What we'd like is a more efficient way to do this and there's something called the binomial expansion, which gives you an automatic way to find things like $(1 + x)^{17}$.

Let me show this to you. It's aimed at figuring out $(1 + x)^r$; I will write it where r could be 17 or 3 or whatever. The expansion looks like this. It starts with 1, when you multiply $1 + 3$ times itself r times, all those 1's will multiply. Then you get an $r \times x$; then you get an $r(r - 1)/2 \times 1 \times x^2$. That's the third term. The x^3 term has its coefficient $r(r - 1)(r - 2)/3 \times 2 \times 1$; that's how many x^3 there are, and the pattern continues. You can see what you're doing here; you always march down from r, $r - 1$, $r - 2$, and so on. In the bottom you're taking 3, 2, 1. The next term would have a 4, 3, 2, 1 in the bottom and so on. It's a very nice pattern. It's called the binomial expansion.

Let's do an example with $r = 2$. Suppose I wanted to go back to my first example $(1 + x)^2$ and not do it by multiplying these out, but by using this binomial expansion. The next term, think about this one. It'll be $2 \times 1 \times 0$ as we descend over $3 \times 2 \times 1$ x^3, but that's going to be 0. The term after that would be $2 \times 1 \times 0 \times -1$ over $4 \times 3 \times 2 \times 1$, but that will be 0 too. In fact all the rest of the terms are gone. They all will have a 0 in them and all we're left with are those first three terms $1 + 2x$ and look at the next one, 2×1 over 2×1 cancel and so you get $1 + 2x + x^2$ as we expected.

Let me use the same expansion for the harder problem $(1 + x)^{17}$. You can see where this is going to earn its keep, $(1 + x)^{17}$; I just substitute in, let r be 17, run across it, you'll get $1 + rx$ is now $17x$ plus $r \times r - 1$ over 2×1, 17×16 over 2×1 x^2, the next term $17 \times 16 \times 15$ over $3 \times 2 \times 1$ x^3, $17 \times 16 \times 15 \times 14$ over $4 \times 3 \times 2 \times 1$ x^4. You can just keep going, the pattern is quite clear and then all you have to do is simplify these fractions. The arithmetic can be a little bit tricky, but it's not bad. If you're paying attention, you'll get $1 + 17x + 136x^2 + 680x^3 + 2380x^4 +$ on and on it goes. You finally grind to a halt with $17x^{16}$ and lastly x^{17}. This is a finite series, not an infinite one, but it's done via the binomial expansion.

Here's one more problem like this with my expansion. What if somebody asked you what's the coefficient of x^{11} in the expansion of $(1 + x)^{17}$? Suppose that's all they wanted to know. If you took $1 + x$ raised it to the 17th power, how many x^{11} would there be? The multiplying out is not what you want to do. The Pascal's triangle is probably not what you want to do. I want to do the binomial expansion, but this time I'm just going to 0 in on the x^{11} term. If you look at the formula, you can see what's going to happen. To get an x^{11} you're going to have a fraction in front of it. That'll be the coefficient. Under it will be $11 \times 10 \times 9 \times 8 \times 7 \times 6 \times 5 \times 5 \times 4 \times 3 \times 2 \times 1$ and on the numerator you'd start with 17 and go down to 16, 15, 14, until you get 11 terms up there. I can immediately write down what the 11th term is going to be, what the x^{11} term is going to be. There it is, this big fraction with all these numbers on top and all these numbers on the bottom. It looks like a mess to try to multiply all this out and simplify it, but actually all kinds of cancellations will occur if you look at that. You see 11, 10, 9, 8, 7 in the top, 11, 10, 9, 8, 7 in the bottom so right away you can lop off most of those terms. What's left isn't too bad; you do the work and you see that there's $12376x^{11}$.

That's kind of background; that was all around before Newton came along. What Newton realized was that we could use this expression in the case of the exponent being non-integer and that's what I want to show you, what you can do if that r isn't 17 or 2. Before doing so, however, let me just introduce a notational convention. This'll just simplify the expression a little bit. When I have something like $4 \times 3 \times 2 \times 1$, what I'm going to write is 4!, which is read 4 "factorial." It just abbreviates the notation so you don't have to write $11 \times 10 \times 9 \times 8$, you just write 11 "factorial." You can always tell when somebody doesn't know math. If they see this symbol like that 4 factorial and they think that means they're supposed to say 4 with a great deal of enthusiasm or passion. That's not an exclamation point that's supposed to require you to get very theatrical. It just has a mathematical meaning.

If I want to go after that, then I can recast my binomial expansion as $(1 + x)^r$ is $1 + rx + r(r - 1)/2! \ x^2$ instead of 2×1 plus $r(r - 1)(r - 2)/3! \ x^3$ and it's in that form I want to continue the discussion. As I say Newton's insight in 1664, while he's still a student at Cambridge, was that you could use this expansion not just for whole numbers r, but for other kinds of exponents. What he said is r could be integral or (so to speak) fractional, positive or negative. He's going to allow fractional exponents up there for r, negative exponents up there for r. Nobody had ever thought of that. He said this is still going to work and it's going to be valuable. Perhaps before we see Newton using this I should just review for a minute what a negative exponent means, what a fractional exponent means.

Real quick, this will be a little review here. What's a negative exponent up to? If I wrote something like x^{-n} what that means is $1/x^n$. The negative exponent means you raise the thing to the n and flip it over, so 3^{-2} is $1/3^2$ or $1/9$. Not everybody in Newton's day would've recognized that. When he wrote about this, he sort of had to stop and explain what this was to people in case they weren't up on it. Likewise a fractional exponent $x^{p/q}$, suppose the exponent is the fraction p/q, what this means is the $(\sqrt[q]{x})^p$ power. When I write $x^{p/q}$, the q the denominator of that fractional exponent is the root you take and then numerator, the p, is the power you raise it to. If I had something like $8^{1/3}$, this means the $\sqrt[3]{8}$ to the first power. You don't need to write the first power, it just means the $\sqrt[3]{8}$, which is 2, so $8^{1/3}$ is 2 and something like $6^{5/2}$ means the second root of 6 to the fifth power. The second root of 6 is the $\sqrt{6}$, so it means the $(\sqrt{6})^5$, i.e. the $\sqrt{6} \times \sqrt{6} \times \sqrt{6} \times \sqrt{6} \times \sqrt{6}$; five of them and any two of them $\sqrt{6} \times \sqrt{6}$ multiplied will give you a 6, so you can see what'll happen here. You'll get 6×6 is 36, $\sqrt{6}$.

With that little background then we will look at the binomial expansion and allow fractional or negative exponents. This is what's called Newton's generalized binomial expansion. He's allowing more general exponents up there than anybody had before. So let me do an example. This is one he looked at. Let's expand the $\sqrt[3]{1 + x}$. Suppose you want to expand this out as a series. You wouldn't actually do it with the cube root; you'd turn that into a fractional exponent. You'd write this as $(1 + x)^{1/3}$, that would be your fraction. When I use my generalized binomial series r, the role of r will be played by $1/3$. Let's do it, $(1 + x)^{1/3}$ will start off with $1 + rx$ will become $1/3(x)$. Next up is plus $r \times r - 1/2!$ x^2, but instead of r we're putting in $1/3$, so we have $1/3(1/3 - 1)/2!$ x^2. Then comes $1/3(1/3 - 1)(1/3 - 2)/3!$ x^3. If you start simplifying that you'll see that you'll end up with $1 + 1/3(x) + 1/3 \times -2/3$ over 2×1 x^2 because $1/3 - 1$ is $-2/3$. The next term will be $1/3 \times (-2/3)$ $(-5/3)/$ $3 \times 2 \times 1$ x^3 because $1/3 - 2$ is $-5/3$ and on it goes. If you then clean up the fractions, watch out for the signs, you'll get $1 + 1/3x - 1/9x^2 + 5/81x^3 - (10/243)x^4$ and so on.

What's critical here is that this series doesn't stop. It never quits. It keeps going; it's an infinite series. Compare this to the one I did earlier where we had $1 + x^2$, remember as those exponents were coming down we had $2 \times 1 \times 0$ and that wiped out all the rest of the terms. Here we never get a 0 in the numerator; we skip over it. I started off with 1/3 and when I subtracted 1 I'm at $-2/3$ and when I subtract again I'm at $-5/3$. You'll never hit the 0. This series never stops and so this is an infinite series. This is the sort of thing for which Newton became justly famous. What I've just shown then is that the $\sqrt[3]{1 + x}$ can be expanded as they say into this infinite series, $1 + 1/3x - 1/9x^2 + 5/81x^3 - 10/243x^4$ on and on it goes. Newton says, "… the extractions of roots are much shortened by this theorem." He thought this was very useful for finding roots. I think I should show you why. What's he mean?

Let's suppose somebody asks you to find the $\sqrt[3]{140}$. Okay so here's one of these challenges. Find the $\sqrt[3]{140}$, no calculators allowed. You've got to do it by hand. Imagine you're back in the 17th century. You know how in the world would you do something like this? You'd use Newton's generalized binomial expansion. However, you've got to be careful here, just one little word of warning. When I use this expansion the value of x that I stick in this infinite series must be small so that the terms zero in on an answer. The modern mathematicians say so that the terms converge to something.

If you look at the series, you can see what happens. I have $1 + 1/3x$, then the next term has an x^2, and the next term has an x^3. If I have a large x that I'm substituting in there and I'm squaring it and cubing it and taking it to higher and higher powers the whole thing could just explode on me. That's not going to be good. That will not be what we want. But, if you put in a small x and take powers it's actually going to start to slow down in its growth. Think about that, think about a 1/2, there's a small x, $(1/2)^2$ is 1/4, smaller, $(1/2)^3$ is an 1/8, smaller.

What Newton realized was the smaller the x you put in here the better. In fact, x has to be smaller than 1 or this thing isn't going to work. That presents a problem to my little calculation of the $\sqrt[3]{140}$. What I would be tempted to do would be to say if I want the $\sqrt[3]{1 + x}$ and I'm trying to do 140 I'll just write the $\sqrt[3]{1} + 139$, $1 + 139$ is 140, I would let x be 139, but that is not going to work. If you put 139 into this series those powers of 139 will just blow you out of the water. The series won't head toward anything; it won't converge to the cube root of 140. A sort of a superficial attack on this isn't going to do it for you. Newton says you have to use artifice, his word, you've got to be tricky. Let me show you the trick, how you do this, getting over this hurdle of having to have a small x.

I'm going to think of the $\sqrt[3]{140}$ and you ask yourself what is a perfect cube that's somewhere in the neighborhood of 140, a perfect cube, some number you can take the cube root of easily that's up there around 140. How about 125; 125 is 5^3. I'm going to write this as the $\sqrt[3]{125}$ plus whatever I need to get up to 140, which is 15. It's kind of weird, you've taken the 140 and broken it into $125 + 15$. Then under that cube root you factor out a 125. I'm going to write this as the $\sqrt[3]{125(1 + 15/125)}$. If you think we're off track here just multiply the 125 back through the parentheses and you'll get $125 + 15$, which is 140. We haven't done anything wrong yet. That expression under the cube root is just another way to write 140, but it seems like a rather peculiar way to do it except for this problem in which it's exactly what you want.

Next up, the cube root of the product is the product of the cube roots. Instead of the $\sqrt[3]{125}$ times that parenthetical we're going to write it as the $\sqrt[3]{125}$ times the cube root of the parenthetical. We break these apart into two cube roots, but the $\sqrt[3]{125}$ is easy. That's 5; I know that, that's why I picked 125. That one I can do. What you're left with beside there is the $\sqrt[3]{1 + 3/25}$. I can use the binomial expansion, now I can use Newton's result because the x is going to be $3/25$, which is small. If I put that into the series, it will converge very quickly to my answer. That's the artifice; that's the little trickery you have to employ. Let's see what happens when we do that. I'm going to get that the $\sqrt[3]{140}$ is $5\sqrt[3]{1 + 3/25}$. I go to the series for the cube root and I'll tell you what I'm only going to take the first four terms and stop. The series goes on forever, but I don't have forever, so I'm just going to take the first four terms and thus I'm going to get an approximation. Sometimes we write a little squiggly equals sign for approximately.

I'm going to end up with 5 times; stick 3/25 into the binomial expansion and you'll get $1 + 1/3$ of x will be a 1/3 of 3/25 minus a 1/9 of x^2 will be minus 1/9 of $(3/25)^2$ plus 5/81 of $(3/25)^3$. I'm stopping there, saying that'll be a good enough approximation. You collect all those terms in the square brackets and it turns out to be 9736/9375 and you're multiplying by 5 on the outside of the square brackets. The 5 cancels into the denominator and you're left with 9736/1875 and if you turn that into a decimal you get 5.19253. That's my approximation of the $\sqrt[3]{140}$ using Newton's result. If I go to a calculator and put in the $\sqrt[3]{140}$, what do I get? I get 5.19249. I'm off by four places in 100,000. That is extremely accurate and I did it with just a bunch of fractions. I could do this by hand. Remember what Newton said, "… the extractions of roots are much shortened by this theorem," as they indeed are.

Apparently he loved this thing so much that he tells us, "I am ashamed to tell how many places I have carried these computations, having no other business at the time, for then I took really too much delight in these inventions." He wasted a lot of time doing cube roots and things. That's one of his great infinite series.

I want to show you another one a little more quickly and it involves the infinite series expansion of the sine. This was from 1669, the sine being a famous trigonometric ratio. I think we better review that, we haven't seen trig yet in this course so let's just remember what the sine is. You have a right triangle, an angle theta that I'll put up there and then sides a, b, and hypotenuse c. We're back in the friendly confines of right triangles. Trigonometry deals with ratios of these sides and in particular the sine of the angle theta is defined to be the length of the opposite side over the hypotenuse.

In my picture, it's b/c. That's the sine of the angle theta and we'll remember that. That's what Newton is going to want to expand. But, I need one other little fact and that is that when measuring angles in problems like this and indeed as we move into more advanced mathematics, we want to measure the angle not in degrees, but in something called radians. The angle will be the same, but it's the units or the procedure by which we measure the size of the angle, the size of the opening that is going to differ. It's sort of like if you had a length you could measure the length in meters, you could measure the length in feet, it's the same length just a different approach to measuring it.

Likewise for an angle we could measure it in degrees. Remember the Greeks would measure these things in terms of how many right angles they had or we could use radians. It turns out radians is the way to go for Newton's series and indeed for calculus and beyond. I want to just tell you what that is, what is a radian measure. What you want to do is find the length of the arc that the angle cuts off on the unit circle if the angle is centered at the center of the circle. That sounds pretty complicated so let me give you a picture of how to measure an angle in radians.

There's my angle theta. It's a certain opening. I will let O be the vertex of the angle and I'll put my compass there and I will then draw a circle of radius 1, so centered at the vertex, circle of radius 1. Let's say that the rays of the angle cut across the circle at A and B then the radian measure of the angle theta, that is the radian measure of angle AOB is just the length of that arc, AB. That's all you've got to do. To measure the angle, you just measure the arc length when the circle has radius 1 and that's called the radian measure.

How about an example? Suppose I need to find the radian measure of an angle of 20 degrees. In our familiar measurement system we say it's a 20-degree angle, I want to convert it to radians, what do I get. Here's how you do it. There's a 20-degree angle, you put O in the vertex, you draw your units circle; that is your circle with radius 1. It hits the rays of the angle at A and B and now my challenge is to find the length of AB. That's going to be the radian measure.

What you say is this. That's not really that hard to do because 20 degrees is 1/18 of the way around the circle. Right 20 degrees over the whole 360 is 1/18. My angle is 1/18 of the way around, so my arc length AB is 1/18 of the circumference around. What's the circumference of this circle? It's $2\pi r$. That's the circumference of any circle except in this case r is 1. This is the unit circle; the radius is 1. The length of arc AB is 1/18 of $2\pi \times 1$, in other words $1/18(2\pi)$, in other words $\pi/9$. An angle with a degree measure 20 degrees has radian measure $\pi/9$—same angle, just a different way of measuring.

Newton using his generalized binomial series and using this newly invented method of his called inverse fluxions—that is, integral calculus developed an infinite series for the sine of x. I'm going to use x instead of theta here, where x is the radian measure of the angle. Here it is. This is Newton's sine series. The sine of x is $x - x^3/3! + x^5/5! - x^7/7!$ and so on. It's a beautiful pattern. Look at that; it's very nice. The odd powers are showing up, x, x^3, 5th, 7th, 9th, the factorials below them in perfect marching order. The signs are flip-flopping plus, minus, plus, minus and if you do this you get the sine of x. This is from Isaac Newton, this famous series, an infinite series. Notice it keeps going forever.

He discovered this in 1669 and put it in his manuscript the *De analysi* as it was called, which wasn't published in 1669. Remember, Newton didn't publish his things very rapidly. It was published decades later, but when it was published this is the page on which the sine expansion appears. If you want to see what this looked like when it finally got in print, he was trying to find the base from the length of the curve given and if you look in section 45 there it says he's trying to find the sine, S-I-N-E of AB. The actual formula that he got is a few lines down where you see $z - z^3/6 + z^5/1/20$ 6 is 3!, 120 is 5!, that's the series using z instead of x. There it is as it appeared in print from Sir Isaac.

Let me just use this. Suppose I have that series for sine x and somebody says find the sine of 20 degrees. We're used to punching this in on a calculator maybe or looking it up in a table, but suppose you're on your own to find the sine of 20 degrees. I mean somebody was on their own originally when they first did this; how do you do it? You go to the series. You first of all say that 20 degrees is $\pi/9$ radians and so you put $x = \pi/9$ into that series. It's an infinite series. I'm only going to use the first three terms. I put in $\pi/9 - (\pi/9)^3/3! + (\pi/9)^5/5!$. I'll stop there. To do the calculation and I get 0.34202 as my estimate of the sine of 20 degrees using Newton's series. What is the sine of 20 degrees on your calculator? That's exactly right, 0.34202. This is splendid. This is how this works.

Newton gave us infinite series; he gave us the generalized binomial expansion that yielded that cube root of $1 + x$ as an infinite series. He gave us the beautiful symmetric sine series. He was a master of this new subject and infinite series expansions would be important forever in mathematics. We'll see them actually popping up in some later lectures. They are a sign that important new exciting mathematics is being done.

However, let me just say, in my next lecture rather than looking to the future as Newton was doing here, I'm going to look to the past. I want to show you one more piece of Newton's mathematics. I want to show you how he proved an old, old result—in fact an old friend of ours from the days of the Greeks—how he proved Heron's formula.

Newton's Proof of Heron's Formula
Lecture 15

The geometry in Newton's argument is much reduced from that of Heron, but there's a price to pay. If the geometry goes down, the algebraic sophistication goes up, and as you'll see, the algebra is where the heavy lifting occurs in Newton's proof. You win some, you lose some.

In this lecture, we look at Newton's proof of the formula by Heron we used earlier to find the area of a triangle and a quadrilateral. As you may recall, for a triangle with three sides, a, b, and c, we find the semiperimeter, s, and ultimately, perform the following calculation: $\sqrt{s(s-a)(s-b)(s-c)}$. Heron's original geometric proof of this implausible result was quite complicated; Newton tried an algebraic approach.

Newton's proof appeared in his textbook on algebra, *Arithmetica Universalis*, published in 1707. At this point in his career, Newton was experiencing a renewed interest in geometry and devoted his time to improving proofs of elementary theorems. He challenged himself to find algebraic proofs of Heron's formula and similar geometric results. For example, one problem he tackled was to find the hypotenuse of a right-angled triangle given the area, a, and perimeter, p. The result? The hypotenuse is $p/2 - 2a/p$.

> **Newton was experiencing a renewed interest in geometry and devoted his time to improving proofs of elementary theorems. He challenged himself to find algebraic proofs of Heron's formula and similar geometric results.**

Newton stated the theorem for Heron's formula as follows: "If from half the collected sum of the sides of a given triangle the sides are individually subtracted, the square root of the continued product of that half and the remainders will be the area of the triangle." Unpacking this statement, we find the equation we've seen earlier for Heron's formula. To prove this result, Newton needed Euclid's proposition 8 from Book VI: Given a right-angled triangle, an altitude drawn to the hypotenuse will split the triangle into two similar right triangles. The similarity of the triangles allows us to build a proportion; for example, long leg is to short leg (x is to h) in one triangle as long leg is to short leg (h is to y) of the second triangle. Cross-multiplying, we get $h^2 = xy$. In other words, if we draw the perpendicular from the right angle to the hypotenuse, the square of that altitude's length is the product of the two parts of the hypotenuse, x and y, into which the altitude splits it.

Newton begins his proof with a triangle ABC. He then constructs a relatively simple figure with two similar triangles, lines extending in opposite directions to points that equal the distance from A to B, and a perpendicular of length h. From here, he applies the Pythagorean theorem to both triangles and sets the two resulting expressions equal: $c^2 - AD^2 = a^2 - DC^2$. The problem is now recognizable as a difference of squares, which can be factored into $(AD + DC)(AD - DC)$. Looking at Newton's construction, we see that $AD + DC$ is the whole base of the original triangle; in other words, it's b. Inserting that in the expression, we get $c^2 - a^2 = b(AD - DC)$, or $c^2 - a^2/b = AD - DC$. The proof continues with a series of steps that account for the various pieces of Newton's construction. In the end, we find that the expressions $2(s - a) = -a + b + c$, $2(s - b) = a - b + c$, and $2(s - c) = a + b - c$ appear in the formula that we've developed—at significant algebraic cost—for h^2. After a great deal of work, we arrive at h^2 (the area of the triangle squared) $= 2(s - c) \times 2(s - b) \times 2s \times 2(s - a)/4b^2$. Simplifying, we get $h^2 = s(s - a)(s - b)(s - c)$; we take the square root to get the area of the triangle and bring to an end this conversation between Newton's algebra and Heron's geometry. ∎

Suggested Reading

Gjertsen, *The Newton Handbook*.

Westfall, *Never at Rest*.

Whiteside, *The Mathematical Works of Isaac Newton*.

1. As noted in the lecture, Newton included the following problem in his *Arithmetica Universalis* of 1707: Find an expression for the hypotenuse (h) of a right triangle whose area is A and whose perimeter is P. Do the algebra to show that the answer is $h = \dfrac{P}{2} - \dfrac{2A}{P}$.

2. Did you find Newton's proof of Heron's formula to be brilliant or tedious? Could you see ahead of time where he was going with his multiple algebraic manipulations? Most people who can follow this step-by-step argument are nonetheless surprised when Heron's formula emerges, seemingly from nowhere, at proof's end. Nonetheless, it is an intriguing "conversation" between two great mathematicians from two different millennia.

Newton's Proof of Heron's Formula
Lecture 15—Transcript

In this lecture, I want to show you Newton's Proof of Heron's Formula. We'd seen Heron's Formula earlier. We used it to find the area of a triangle and of a quadrilateral, but we didn't prove it. At that time I said we'll defer the proof until later; later has arrived. I should say that this particular argument is one that's laden with algebra. There'll be formulas marching across the page in great abundance so, as always, I'll try to go very slowly and very carefully in order to derive this formula.

Nothing here is harder than high school math; it's just high school math with a vengeance so let's see what we can do with this great proof. First, I'd better remember what Heron's Formula is. Remember, it was a way to find the area of a triangle. You have a triangle with sides A, B, and C. Remember, the formula says you first take the semi-perimeter. You take A + B + C, the perimeter, divide it in half, and then the area of the triangle is the square root of that—S times S – A times S – B times S – C.

This is an extremely implausible result. We really need to prove this so you believe it it's such a strange formula for triangular area. Heron's original proof, which is dated at around 75, was quite complicated. That's why I didn't do his proof. He started, of course, with the triangle whose area you need. The first thing he did was inscribe a circle within in. To inscribe a circle means the circle is tangent to the three sides of the triangle. That seems like a strange way to begin, but that's how he did it. How do you know you can inscribe a circle in a triangle? It's Euclid, Book 4.

He then used properties about cyclic quadrilaterals. We've met these before. A quadrilateral is cyclic if it can be inscribed in a circle. Heron needed special properties of these things, which we would have to have stopped and done separately had we gone down Heron's proof. He needed to introduce three pairs of similar triangles. When you take it all together, it was quite a sophisticated geometrical argument. Here's his picture; this is the proof that Heron uses.

You can see the triangle ABC. You see he put a circle there, inscribed in the middle. He drew all sorts of lines all over the place so that it's quite cluttered with lines. I don't know if you see it, but there's actually a cyclic quadrilateral there someplace—a quadrilateral in a circle. It's AHCO and there's the circle that goes through the points. He put all this together. It was a beautiful proof, but kind of sophisticated so that's not the way we're doing it. We're going to go to Newton's proof. His proof of the same result is much more algebraic, much less geometric.

He starts with a triangle of course, as does Heron, no need for an inscribed circle. That's good. No need for cyclic quadrilaterals, there aren't any in Newton's proof and only one pair of similar triangles and they're easy. The geometry in Newton's argument is much reduced from that of Heron, but there's a price to pay. If the geometry goes down the algebraic sophistication goes up and as you'll see the algebra is where the heavy lifting occurs in Newton's proof. You win some, you lose some.

The proof itself appeared in Newton's book the *Arithmetica Universalis* published in 1707. This was his textbook on algebra. It was actually his bestselling book in his lifetime. His other more sophisticated books didn't actually see very well, but this one did. It was more elementary. He promised in the preface to, "Let the Learner proceed to exercise himself ... by bringing Problems to Equations ..." We're going to give you a problem and we're going to bring it to an equation and then employ the techniques of algebra.

Here's a problem that he gave in this *Arithmetica Universalis*. He said if the number of Oxen a eat up the Meadow b in the Time c; and the Number of Oxen d eat up as good a Piece of Pasture e in the Time f, and the Grass grows uniformly; to find how many Oxen will eat up the like Pasture g in the time h. This is a horrible problem. This is the sort of thing that gives students nightmares. It's like fingernails down the blackboard. I don't know why Isaac Newton put this in his book; it doesn't excite anybody I don't think. He has other things in the book that are better than this, but I just wanted to show you that horrible algebra problems go all the way back. If you want the answer of the Number of Oxen is given by that horrible formula, which I'm not even going to read.

But, if that problem wasn't particularly successful, others were pretty good. Derek Whiteside says that Newton turned his attention to geometry here and that his proof of Heron's formula, which is in this book, occurred when Newton experienced "a renewed interest in pure geometry." At this point in his career he was looking back at geometry and he devoted himself to "constructing improved proofs of … elementary theorems."

What Newton set before him as a challenge was to go find elementary theorem like Heron's formula or some other geometrical result and give a nice proof of it using algebra. This was supposed to be the book that showed the value of algebra. Newton observed this in the book. He says, "Geometrical Questions may be reduced sometimes to Equations …" Ahh a geometrical problem can be turned into an equation, an algebra problem. "But," he warned "in Geometrical Affairs … they so much depend on the various Positions and complex Relations of Lines, that they require some farther Invention and Artifice to bring them into Algebraick Terms."

Geometry problems are challenging to turn into algebra problems, he says, because of the way the lines stack up, so you have to be careful. You need some artifice, but boy Newton was great with his artifice, so he could do this. Let me show you one geometry problem before we get to the Heron. He says, the area and perimeter of a right-angled triangle being given, to find the hypotenuse. Here was a little problem that sounds geometrical, but he's going to make it algebraic. Suppose we're given a right triangle whose area is given, we'll call it A, whose perimeter is given, we'll call it P; challenge find h, the hypotenuse. There's the triangle; I supposedly know its area. I know its perimeter; I've got to find its hypotenuse.

This is a good problem to try at home. See if you can solve one of Isaac's problems. He tells the answer; I'll tell you what it is. The hypotenuse is P/2 – 2A/P, half the perimeter minus twice the area over the perimeter. It's a nice exercise. He did other things here. He did things like problem 19 in the book is he wants to show you how to surround a fishpond with a walk of given area and the same breadth everywhere. Here's a geometry problem that again is going to turn into an algebra problem. Imagine you have a fishpond and you have a given amount of asphalt let's say. That's that given area. You only have so many square feet of asphalt. You want to surround the pond with a walk that's the same width all the way around. How do you do it? He shows how to use algebra to solve that problem. He's doing geometry via algebra. But, the problem that's particularly important for me is this one problem 12 where he says having the sides and base of any triangle given to find the area.

Give me a triangle ABC, I give you the sides, a, b, and c—what's the area? If you look at this, you'll see what's up here. He's given you the three sides of the triangle and wants the area, what's he after? He's after Heron's formula. Here is his page of scratch work. Here is where he did this. This is actually Newton writing the proof. You can see his handwriting. You can see the crossings out. Let me just call your attention to a few things on this page from his notes.

In the upper right you see his diagram, the little triangle he's going to use to prove this. I will reproduce this in better form on the screen, but notice it's much simpler than that diagram that I showed you from Heron that Newton got away with a much simpler picture. Notice also and perhaps you can see this, at the very top on the upper left it says Propositio, so in Latin he's saying here comes the proposition. A few lines down it says Constructio—ahh, here comes the constructions he's going to need to do his proof. A little further down yet it says Demonstratio, the demonstration, the proof that it works, the proof that Heron's formula is true. It's all set up ready to go on that one page from his notes.

Here's how he states the theorem. Newton says, "If from half the collected sum of the sides of a given triangle the sides are individually subtracted, the square root of the continued product of that half and the remainders will be the area of the triangle." It's a most strange way to state Heron's formula. I don't know why he didn't just write down the formula. He was trying to extol the virtues of algebra it would seem like a great place to give this as an algebraic expression, but he chose to write it in this prosaic form. It's kind of neat though and if you look at you see what he's up to.

He says to find the area of a triangle you take half the collected sum of the sides. Well the sum of the sides is the perimeter, half the collected sum; he's talking about the semiperimeter there. He says suppose the sides of the triangle are individually subtracted from this. We're going to take $s - a$, $s - b$, $s - c$, subtracting the sides from this semiperimeter. Then he says the square root of the continued product of that half and the remainders will be the area of the triangle. The product of that half s and those remainders $s - b$, $s - c$, $s - a$, that product, take the square root you've got the area of the triangle. He's put it in a rather strange way, but that's it if we can prove it.

One more result before we plunge in and this is something we saw in an earlier lecture. This is from Euclid Proposition 8 of Book 6. We saw that if you have a right-angled triangle ABC and you draw the altitude down to the hypotenuse it splits the right triangle into two right triangles that are similar to one another. We did this. The green triangle is similar to the blue in the picture and we're going to need this. Newton's going to need this particularly in this sense.

If I call the length of the altitude h, that perpendicular h, and it breaks the hypotenuse AC into two pieces, one of which is x and one of which is y, then the similarity of the triangles will allow me to build a proportion here. For the green triangle I can say long leg is to short leg that is x is to h, as in the blue triangle long leg is to short leg, h is to y. If you cross multiply this you get that h^2 is xy. When you draw the perpendicular from the right angle to the hypotenuse, the square of that altitude's length is the product of the two pieces of the hypotenuse, x and y into which the altitude splits it. That we'll need as well. Now we're ready. Here goes Newton's proof. Here comes the algebra. We will take a deep breath. There's the triangle ABC, sides a, b, and c, our object is to show that its area satisfies the wonderful result of Heron. Newton says first thing bisect AC at M. So AC, we'll put the longest side AC down there along the bottom and I split it in half at M. You can do this with a compass and straightedge so there it is. From A to M is $b/2$ and from M to C is $b/2$ because AC that whole side there was of length b.

Next he says on CA extended construct AF equal to c. I take that line CA and I keep it going off to the left and I go as far from A to a point F so that AF is c. What was c? The c was that side of the triangle from A to B. What I'm saying is you'll just take CA and go out exactly as far to F as it is from A to B. Next up he says go the other way. Construct AE rightward from A so that AE is also equal to c. In my picture I'm going to go to the right of A to some point E so that AE is c in that direction. At this point we've gone from A leftward to F c units and rightward to E c units, c being the distance from A to B. So he's building an interesting looking picture here.

There are a couple of more steps in his construction. He says draw BF and BE, so we draw the line from B to F, there it is, the line from B to E, and drop BD perpendicular to AC of length h. We're going to draw up the perpendicular from B straight down to AC, there it is of length h. That's all he needs to do. That's his whole picture. That's the construction. It's pretty simple, it's just a matter now of finding our way through the thicket that will appear here in terms of all the algebra that will arise.

Let's begin. The first thing Newton says is look at that triangle ADB. I've turned it into a blue triangle. That's a right triangle because we drew BD perpendicular, so that's a right angle, that's a right triangle, we can invoke our old friend the Pythagorean theorem, and conclude that c^2 is equal to h^2 plus the horizontal leg AD^2. Therefore, c^2 is h^2 plus AD^2 and actually I want to write it this way $c^2 - AD^2$ is the h^2, the Pythagorean theorem sort of moved around a bit.

Now look on the other side to the right of that altitude I have a red triangle. It also is a right triangle, the Pythagorean theorem applies just as well to it saying that a^2 is h^2 plus DC^2 or if I isolate the h^2, h^2 is $a^2 - DC^2$. You see the h is common to both triangles. The h^2 shows up in both Pythagorean Theorems and I isolated and set them equal. Now $c^2 - AD^2$ on the far left equals $a^2 - DC^2$ on the far right and I move things around to get $x^2 - a^2$ is $AD^2 - DC^2$. Now what happens? Look at that thing on the right side, $AD^2 - DC^2$, that looks familiar. That's called a difference of squares. You see this in algebra. We've actually seen it in this course already, $x^2 - y^2$ and you have a difference of two squares and you can always factor that into $(x + y)(x - y)$, basic high school algebra. That's just what I'm going to do with that $AD^2 - DC^2$. Let me look at that, the difference of squares; that will factor into $(AD + DC)(AD - DC)$.

Now you look at the picture and see what you can identify here. What is AD + DC? Look at the picture, from A to D and then D to C is like going from A to C. It's the whole base of the triangle. In other words it's b. Where AD

+ DC was, I'll put b, where AD – DC is I'll just leave that there for now. We see that $c^2 - a^2$ is equal to b times this expression (AD – DC) and I'll solve for that. I'll get $c^2 - a^2/b$ is AD – DC. That's one of the steps along the way in this algebraic journey we're taking.

There it is, $c^2 - a^2/b$ is AD – DC, that's where we are. Newton is going to look at these two expressions, this AD and this DC and see what we can make of that. Look at AD, look at the picture. That's just (AM + MD). If I'm going to go from A to D I could just as well go from A to M and then the rest of the way from M to D, sure. Look at DC, the other part of this expression. DC is the same as MC – MD. Again you can see right there on the triangle. If I went from M to C, but then subtracted the piece from M to D I'm left with DC. There's where I am and now something important here. M, remember, was the midpoint of AC. So what is AM? AM is $b/2$, half the base. What is MC? That's $b/2$ also, half the base. We stick those variables in and we get as my expression $(b/2 + MD) - (b/2 - MD)$. The $b/2$'s will cancel and you'll have two MDs left here. So $c^2 - a^2/b$ is 2(MD), therefore MD is $c^2 - a^2/2b$ and that's another step along the way.

The trouble is that isn't the real important piece he's after. MD isn't the real issue. The real issue believe it or not from this picture is to figure out DE. That comes next, so here's how he's going to do that. He's going to say what is DE, how long is that? Again looking at the diagram we can see that it has to do with other pieces or other segments. Isn't it just ME – MD. Look up there, ME – MD, that would leave me with D. What's ME? Newton says that is (AE – AM). We look at the picture once more. AE – AM; that's ME. Meanwhile we still have the –MD there tacked on the end, so we have to attack this thing. But, now all the pieces are finally going to become clear. AE if you look at that was c. Remember that's how we did that, we constructed from A over to E c units, so where AE is c's going in. AM is half of the base, that's $b/2$, and MD is this expression we figured out previously $c^2 - a^2/2b$. Here comes the algebra. Everything goes in there. We get $(c - b/2) - c^2 - a^2/2b$, a common denominator would be called for, how about that would be $2b$ and thus DE it turns out is $2bc - b^2 - c^2 + a^2/2b$, the result of combining all those fractions over the denominator 2b.

Now we're going to store that result for DE for just one minute, we will return, but I have one more observation here. Newton says triangle FBE is a right triangle. Triangle FBE, the ones with the dotted lines on it, FB and BE, that's a right triangle—why is that? Here's why. Suppose you took your compass and put the center at A and had your radius C and drew a circle. From A to F is c units, so F would be on the circle. From A to B is c units, so B would be on the circle. From A to E is c units, so E would be as well. If I draw in the circle, you see that angle FBE is an angle inscribed in my semicircle.

But, what do we know about an angle inscribed in a semicircle? It's a right angle and we know that from Thales, so we go back to our original proof, one of the early theorems in all of mathematics is just what we need here, Thale's theorem. Remember what I said in my initial lecture. I gave that old quotation that, "Logic can be patient because it is eternal." Thale's proof is eternal. We can still use it. Triangle FBE is a right triangle. Therefore, BD is an altitude to the base to the hypotenuse FE. I can use the similarity argument from Euclid to conclude that FD, that piece of the hypotenuse, is to h as h is to DE, the other piece of the hypotenuse, and hence by cross multiplying h^2 is DE × FD.

But, FD, we want to break it up; this is our last break up I promise. FD if you look at the picture is FE – DE, FE – DE, and so h^2 is DE(FE – DE)—but FE, we can see what that is. From F to A is c, from A to E is c, FE is $2c$ and so I reach this conclusion that h^2 is DE times the quantity $2c$ – DE. There it is. But, remember we knew what DE was. We had found that a couple of steps ago. DE is $2bc - b^2 - c^2 + a^2/2b$. Where the DE is I put in that big monstrous fraction. Then come the parentheses and I have $2c$ – DE, so it's $2c$ minus that big monstrous fraction, $2bc - b^2 - c^2 + a^2/2b$.

The first expression's fine, I'll just let that be, but the second expression, the one in the parentheses, I need a common denominator there to put that together into a single fraction. The denominator's $2b$; that'll require a $4bc$ to show up in the front end. Then you get minus the $2bc + b^2 + c^2 - a^2$ and so this is what h^2 is. If you combine the terms in that second numerator, you'll just get $2bc + b^2 + c^2 - a^2$. That's what h^2 is. The denominators aren't bad, but the numerators are in need of repair, in need of work.

Let me look at that left-hand numerator. It was $2bc - b^2 - c^2 + a^2$. We can fiddle with this a bit and write it as $a^2 - (b^2 - 2bc + c^2)$. If you check the minus signs through the parentheses you'll see that's still equal. But, that thing in the parentheses should look familiar. That is $(b - c)^2$, $(b - c)^2$ is $b^2 - 2bc + c^2$. And so I have $a^2 - (b - c)^2$. That's a difference of squares again. You see it, it's that same structure $x^2 - y^2$ and you can factor it into $(x + y)(x - y)$. I can factor $a^2 - (b - c)^2$ into $a + b - c$ and $a - (b - c)$. In other words it'll come out to be $(a + b - c)(a - b + c)$.

Back I go to h^2, $DE(2c - DE)$. We had those two fractions; I've just cleaned up the numerator of the left-hand one. The right-hand one, $2bc + b^2 + c^2 - a^2$, that numerator can also be repaired. The first fraction we said broke down into $(a + b - c)(a - b + c)/2b$. The right-hand numerator. If you look at that, the $2bc + b^2 + c^2$, those first three terms are just $(b + c)^2$. The numerator on the right is $(b + c)^2 - a^2$, lo and behold the difference of squares again. If I can put all of this together into one gigantic fraction, the bottom will be $4b^2$ when I multiply the two b's together. Up on the top I have $(a + b - c)(a - b + c)$ and then I employ the difference of squares for that upper right-hand expression and I get $(b + c + a)(b + c - a)$, that's h^2.

We're almost there except it seems like something's missing here, the semiperimeter. We're trying to get something about Heron's formula. I don't see a semiperimeter anywhere; here it comes. Remember the semiperimeter was $a + b + c/2$. If I cross multiply that twice the semiperimeter, that is $2s$ is $a + b + c$. But, look at my expression for h^2. One of those terms, the third one in the numerator, is $a + b + c$. That's going to become $2s$. How about $s - a$, what's that going to be? The s is $a + b + c/2$, the semiperimeter, minus a. Put that into a fraction and you get $-a + b + c/2$, cross multiply and $2(s - a)$ is $-a + b + c$. There it is; it's in that numerator for h^2. Likewise you can show that $2(s - b)$ is $a - b + c$ and $2(s - c)$ is $a + b - c$. All of those expressions are showing up in my formula that I've developed at great cost for h^2. Here we go. Let's revisit it; h^2 we said has a $4b^2$ in the bottom. Up on top there was an $a + b - c$, but I'm going to replace that with $2(s - c)$, that's what that is. Next up in the numerator was an $a - b + c$, but that's $2(s - b)$. Next up was a $b + c + a$, that's $2s$ and lastly $b + c - a$ is $2(s - a)$. There's what h^2 is in terms of these semiperimeters.

We're almost home because there's my triangle ABC; let me look at the area of the triangle squared. I'll drop the h, the altitude. The area of the triangle squared is $[1/2bh]^2$. I used the old formula for triangular area $[1/2bh]$. If you square that you'll get $b^2/4$, when you square the $1/2b$ times h^2. But, after all this work, I finally know what h^2 is, h^2 is $2(s-c) \times 2(s-b) \times 2s \times 2(s-a)/4b^2$. The triangles area squared is that expression. Now comes the exciting drum roll. The b^2 on the top cancels the b^2 on the bottom. In the numerator I see $2 \times 2 \times 2 \times 2$, 16 and the bottom 4×4, 16, all that cancels and you get just $s(s-a)(s-b)(s-c)$, but that's the area of the triangle squared. If the area of the triangle squared is $s(s-a)(s-b)(s-c)$, it follows that the area of the triangle is the square root of $s(s-a)(s-b)(s-c)$. That is Newton's algebraic proof of Heron's formula.

I think that's a wonderful argument, it's a lot of work. It's very clever; it's kind of a conversation between the great Newton and Heron from so long before. It reminds us of why Isaac Newton belongs on the mathematical Mt. Rushmore.

The Legacy of Leibniz
Lecture 16

Remember Newton had his *anni mirabilis*, his miraculous years, when he's basically charting the course of modern science. Leibniz's miraculous years were in Paris, and they were miraculous indeed.

Gottfried Wilhelm Leibniz, born in Leipzig in 1646, was almost an exact contemporary of Newton and has been described as a universal genius. As a boy, the young Leibniz was obviously brilliant. He finished his undergraduate degree by about age 16 and had a doctorate in law by age 20. Unlike Newton, who spent at least the first part of his career as a professor at Cambridge, Leibniz spent all of his adult life as a public servant, but the extent of his interests and talents is breathtaking.

Leibniz, along with Descartes and Spinoza, is considered to be one of the three big names in 17th-century philosophy. He studied symbolic logic and worked on the binary system, which as we know, would become critical in the modern computer age. He was also a student of history, political theory, philology, physics, and engineering. In 1672, Leibniz was sent on a diplomatic mission to Paris, where he found a mentor, Christiaan Huygens, who could guide him in his mathematical training. Huygens was a Dutch mathematician and scientist who had invented the pendulum clock and discovered the rings of Saturn. Huygens challenged the young Leibniz to find the sum of the reciprocals of the triangular numbers: $1 + 1/3 + 1/6 + 1/10 + 1/15$ … ? Leibniz discovered the finite answer to this infinite series, 2, and Huygens agreed to mentor him in mathematics.

Leibniz plunged into mathematical studies and, with his characteristic zeal, raced to the frontier of knowledge.

Leibniz plunged into mathematical studies and, with his characteristic zeal, raced to the frontier of knowledge. In 1674, he solved another infinite series that is now known as the Leibniz series: $1 - 1/3 + 1/5 - 1/7 + 1/9 - 1/11$…. The value of this sum is $\pi/4$. In 1673 and 1676, Leibniz made diplomatic trips to London and was made a member of the Royal Society. During the later visit, he saw a manuscript of Newton's *De analysi*, and he later wrote to Newton, introducing himself and asking about Newton's work in mathematics. The two geniuses corresponded briefly.

In 1684, Leibniz consolidated his ideas on calculus and published the first paper on the subject, entitled "*A Nova Methodus*." For Leibniz, calculus was a set of rules for doing maximum and minimum problems, as well as tangents. In his 1684 paper, he stated the differential rules for sums, differences, products, and quotients. Two years later, he published the first paper on integral calculus.

During the last years of his life, Leibniz became involved in a dispute with Newton over who created calculus, a subject we will cover in greater detail in the next lecture.

The Teaching Company Collection.

The genius Leibniz earned his Ph.D. by the time he was twenty years old.

Leibniz died in 1716, and unlike Newton's, his funeral was a small affair. He is, however, immortalized in a statue at the University of Leipzig. ■

Suggested Reading

Boyer, *The Concepts of the Calculus*.

Child, ed., *The Early Mathematical Manuscripts of Leibniz*.

Dunham, *The Calculus Gallery*.

Edwards, *The Historical Development of the Calculus*.

Grattan-Guinness, *From the Calculus to Set Theory*.

Hall, *Philosophers at War*.

Hoffman, *Leibniz in Paris*.

Newton, *The Correspondence of Isaac Newton*.

Questions to Consider

1. We saw how Leibniz cleverly summed the infinite series of reciprocals of the triangular numbers to get $1 + \frac{1}{3} + \frac{1}{6} + \frac{1}{10} + \cdots = 2$. But he didn't stop there. He also summed the reciprocals of the so-called "pyramidal numbers," which are those with denominators of the form $\frac{n(n+1)(n+2)}{6}$. Thus, the infinite series in question is $P = 1 + \frac{1}{4} + \frac{1}{10} + \frac{1}{20} + \frac{1}{35} + \frac{1}{56} + \cdots$. Here's how Leibniz determined its value: First note that $\frac{2}{3}P = \frac{2}{3} + \frac{2}{12} + \frac{2}{30} + \frac{2}{60} + \frac{2}{105} + \frac{2}{168} + \cdots$. Now rewrite $\frac{2}{3} = 1 - \frac{1}{3}$, $\frac{2}{12} = \frac{1}{3} - \frac{1}{6}$, $\frac{2}{30} = \frac{1}{6} - \frac{1}{10}$, and so on and thereby show that $P = \frac{3}{2}$ *exactly*.

2. For those with a philosophical bent, review Leibniz's contributions to this field. You might also read up on how Leibniz's philosophy of optimism was satirized via the Dr. Pangloss character in Voltaire's *Candide*. There, Pangloss contends—in true optimist fashion—that ours is the "best of all possible worlds."

The Legacy of Leibniz
Lecture 16—Transcript

We've spent time meeting Isaac Newton, this intellectual force of nature from the 17th century. You would be forgiven for thinking that there would be no one, no one that could go toe to toe with Sir Isaac on intellectual matters, but you would be wrong. This is because he had a contemporary who lived across the English channel in Europe who was his peer. His name was Gottfried Wilhelm Leibniz and that's the subject of this lecture.

This is Leibniz. You recall I've shown you a portrait of Newton called *Newton in His Own Hair*. This I call *Leibniz in Somebody Else's Hair*. He's wearing a very splendid wig as was the fashion of the time. He was born four years after Newton and died a little bit before, but almost exactly a contemporary. He was born in Leipzig in 1646 to a scholarly family. Leibniz's father died when Leibniz was just a boy. His father possessed a wonderful library and perhaps a little surprisingly, they let the little boy, little Gottfried, roam free in the library. Leibniz grew up while other kids were out playing, he was reading books from his father's library. He made enormous progress and developed eclectic tastes while grazing through those volumes.

By 1658—when he was 12—he had taught himself Latin. He was writing poetry in Latin and, for good measure, he taught himself Greek as well. He was obviously brilliant. In 1662, he finished his undergraduate degree. At age 20, he had his doctorate in law. His training, thus, was in jurisprudence and his career was not in academe. Unlike Newton, who spent at least the first part of his career at Cambridge as a professor, Leibniz never did that. He was a public servant, which, of course, is what Newton was at the end of his day.

Leibniz has been described as a universal genius. He was somebody who seemed to be able to master anything, to shoot to the frontier of any subject. It's really breathtaking, the sweep of his interests and talents. He studied philosophy. In fact, most people today, if they know the name Leibniz, probably would say oh, he was a great philosopher. He was a great philosopher. It's been said that from the 17th century the big names in philosophy are Descartes, Spinoza, and Leibniz. He certainly was a very important philosopher although we know that he was mainly a mathematician.

He studied logic. He was interested in symbolic logic and was always trying to refine his notation, his symbolism, trying to convert human discourse into symbols so that an argument forum could be checked, for instance, algebraically. He also worked on the binary system and although he couldn't have known it, this of course would become critical in the modern computer age.

He was a student of history, a great historian, and as we'll see late in his life, he got entangled in historical studies that essentially robbed him of much of his free time. He studied political theory; mainly he was a diplomat for much of his life working on different courts, trying to affect compromises. He wanted to bring back together the Protestant and Catholic pieces of the Christian church. He didn't succeed in that, but he thought that was a worthy goal. He studied philology, which is the study of language and he was a master of languages. It was said that Leibniz was the greatest Sanskrit scholar of his time. He studied physics, made contributions to that. He studied engineering, applied physics, and he was a great fan of Chinese culture where he was again one of the masters of the subject.

We live in an era of specialists. It seems to me that Leibniz's specialty was omniscience. He seemed to know everything. He did this in part by voraciously reading and it was said that he was so avid a reader that he "bestowed the honor of reading on a great mass of bad books." He'd just read everything, good or bad. What's missing from the list, however, is mathematics. You notice I didn't say he was, at this point at least, a great mathematician. But, that's coming.

In 1672, Leibniz is sent on a diplomatic mission to Paris. He was working at the time for one of the kingdoms of Germany. Remember Germany back then was not a single unified state, but a bunch of little entities, each of which had its own diplomatic core, Leibniz goes off to Paris representing one of these states. This was a wonderful place to be in 1672. This was Paris under Louis XIV, Paris under the Sun King. Scholars from all over the world would gravitate to this great city. There were musicians, there were writers, there were scientists, there were artists, and here comes Leibniz whose interests are so vast he was never happier than during his diplomatic years in Paris.

However, he noted that he had this gap at this point in mathematics. He wrote this: "When I arrived in Paris in the year 1672, I was self-taught as regards geometry, and indeed had little knowledge of the subject, for which I had not the patience to read through the long series of proofs …" He knew he was lacking mathematically. He tried to read Descartes' *Geometry*, but he found it difficult, dense, and hard to follow. Remember Newton tried to read Descartes and he found it difficult. Descartes has the honor, the privilege, of having stumped two of the greatest geniuses of all time, young Newton and young Leibniz. I don't know if that's an honor he would want, but in any event Leibniz said, "Nevertheless, it seemed to me, I do not know by what rash confidence in my own ability, that I might become the equal of these if I so desired." If I really worked at it, he said, I think I could master Descartes. I think I could master mathematics. He knew something about his abilities. It was just a matter of getting the right direction and jumping in.

What he wanted to do is find a mentor, somebody that could guide him in his mathematical training. It turned out in Paris at the time was just the person, Christiaan Huygens. Huygens was a Dutch mathematician and scientist. He was in Paris like so many mathematicians and scientists and was quite famous. Huygens had invented the pendulum clock, the clock that works with a pendulum and has a whole lot of mathematics behind it. He had also discovered the Rings of Saturn. Huygens was the first person to explain what these strange appendages were that appeared through the telescope when we look at Saturn.

He was a star and Leibniz goes to see him. You can sort of imagine the meeting—the young Leibniz, not very well trained, knocking on the door of the great Huygens and saying, Professor Huygens, will you guide me in my mathematical training? I imagine Professor Huygens said who are you, who is this guy? He's some diplomat; he's not a great mathematician—why should I waste my time on him? What Huygens does is he says, young man, I will give you a challenge. I'll give you a problem and if you can do it that would show me that you're worthy of my time and I can mentor you and lead you into mathematics, but perhaps Huygens figured he'd never see the guy again because the problem was too hard.

Let me show you what the problem was. This was the challenge problem to Leibniz from Huygens. Huygens said we know there are these things called the triangular numbers, these are numbers that are shaped, you can put them into a triangular form, let's put it that way. The square numbers can be put into little squares, 4 is 2 and 2, 9, 3, 3, 3, but you can also build triangular numbers. The first of these is 1, just think of that as a kind of elementary triangle and then 3, you can have 1, 2, 3, one at the top, two below. The next one up is a 1 and a 2 and a 3 or a 6 as a triangle. The next one up is 1, 2, 3, 4, 10. If you've ever gone bowling you should recognize that shape. The next one will be 15 and so on. These are called the triangular numbers.

Suppose I ask for their sum. What would happen if you added up $1 + 3 + 6 + 10$? Obviously that would explode to infinity. You'd be adding up numbers that are getting ever larger; that would be an infinite series that explodes and becomes infinite. That wasn't the problem. What Huygens said was what if we take their reciprocals, take the triangular numbers and flip them over, so look at 1/1, which is $1 + 1/3 + 1/6 + 1/10 + 1/15$, now there's an infinite series that might actually have a finite answer. What is it, young man? What's that sum? Leibniz is sent away having to figure out the sum of this infinite series.

I should say that Huygens knew the answer. This wasn't cutting edge research, but infinite series remember at this time were still not widely known and you couldn't just look this up in a book somewhere. If you were going to figure this out, you were going to have to do some work. But, Leibniz knows no mathematics; he's got nothing to go on except this extraordinary IQ. Here's what he does. He says I will figure out what this is as follows. I'm going to let S stand for the sum I'm seeking. We'll just call it something for the time being. S is this sum $1 + 1/3 + 1/6 + 1/10 + 1/15$, this infinite series whose value we want. If I can just figure out S I'm in business.

Leibniz looked at this; you imagine he thought about it awhile. He says here's what I'll do. Multiply everything by 1/2, both sides. On the one side I'll get 1/2S, on the left. On the other side I march through the fractions and hit each of them with 1/2 so the $1 \times 1/2$ is 1/2, the $1/3 \times 1/2$ is 1/6, the $1/6 \times 1/2$ is 1/12, the $1/10 \times 1/2$ is 1/20 and so on. There's the series for 1/2 of S, which on the surface doesn't look any easier than the one we started with, but it actually is because Leibniz says look here's what I'm going to do. You see that 1/2 on the right side of the equals sign. I'm not going to write it that way. I'm going to write it as $(1 - 1/2)$. Sure, 1/2 is $1 - 1/2$—that's legal, you can do that. You know why he's doing it, hang on, but it's fine. The next number beside it is 1/6, now 1/6 is $(1/2 - 1/3)$. You can see that by getting a common denominator and $(1/2 - 1/3)$ is surely 1/6. The next number on the right is 1/12, that's a $(1/3 - 1/4)$ and 1/20 is a $(1/4 - 1/5)$ and so on. Leibniz looks at this series of numbers and something spectacular happens. You have $(1 - 1/2)$ and then plus 1/2, they cancel, minus 1/3 and plus 1/3, gone, minus 1/4 and plus 1/4, gone, minus 1/5 and plus 1/5, everything's gone except that 1 at the front end.

What he could conclude is that 1/2S is 1. But if 1/2S is 1 then S is 2 and S is what he sought. S was the sum of the series, S is 2. In other words $1 + 1/3 + 1/6 + 1/10 + 1/15$, the sum of the reciprocals of the triangular numbers, is 2. Leibniz takes this solution back to Huygens and says 2, and Huygens says I'll help you. Huygens was impressed. Leibniz passed the test. Huygens becomes Leibniz's mentor. Huygens would say you've got to read this; you've got to read Pascal. Go back and read Descartes, try to make some progress. He wasn't his teacher in a formal sense anymore than Isaac Barrow had been Newton's teacher in a formal sense, and yet they both served as guides, the seasoned older mentors.

Leibniz plunges in and with his characteristic zeal races to the frontier of knowledge. He tells us that he was now "... ready to get along without help, for I read [mathematics] almost as one reads tales of romance." Reading math for Leibniz was as easy as romance novels. He made extraordinary progress during this period. Remember Newton had his *Anni Mirabiles*, his miraculous years, when he's basically charting the course of modern science. Leibniz's miraculous years were in Paris and they were miraculous indeed.

Here's one of the diagrams he created for one of his proofs at this time and I just show it to you; it's kind of a pretty picture. There's a circle; there's arcs flying all over the place. What's he doing with this? He's figuring out what is now known as the Leibniz series. It was 1674 and using that diagram and techniques which we can now see as calculus he figured out this series, another infinite series.

Here comes another one of these. This one very simple in its structure $1 - 1/3 + 1/5 - 1/7 + 1/9 - 1/11$; do you see what it is? The odd numbers in the denominator and the signs flip-flopping, very, very easily grasp what the next term would be for instance. No question there. But, what's the value of the sum of this thing if you were to go on forever? That's not at all obvious, but Leibniz figures it out. The answer's a real shocker. The answer is $\pi/4$. Now $\pi/4$, where did the Pi come from? Where's the circle I always expect Pi to be affiliated with circles. I don't see a circle. I just see a bunch of reciprocals of fractions, but this is right and this is called the Leibniz series in his honor. Leibniz reports that,"[Huygens] praised it very highly and said that it would be a discovery always to be remembered among mathematicians." We still remember it today. Huygens was right.

In 1673 and in 1676 Leibniz made diplomatic trips to London. He was sent from Paris to London to do something for the monarch who was paying his salary. While there he took with him a calculating machine that he had invented that added, subtracted, multiplied, and divided. We had seen that Pascal had invented a machine that could add and subtract; this was the first four-function calculator if you will. It was very cumbersome. It was made out of various components that didn't quite fit right and Leibniz had a devil of a time getting it to work, getting it to grind the gears and crank the levers, but it at least in theory worked. The people in London were very impressed and they made Leibniz a member of the Royal Society in London. Remember, Newton was a member of the Royal Society as well, although at this point he was up in Cambridge rather than being in London.

In 1676, on that visit to London, Leibniz saw a manuscript of Newton's *De analysi*. I mentioned this in the previous lecture; this is where Newton's series for the sine of x appeared. Newton didn't publish this; remember he was not publishing his works. He would write them up very neatly, but occasionally he would loan them out. Somebody would say you know Professor Newton can you show me what you're doing and Newton would say here you can look at this for awhile. You may not publish it, I want it back, but it just so happened that when Leibniz was in London, so was Newton's *De analysi*. Leibniz asked the British mathematicians what's hot in mathematics here in England? Someone said you should look at this, Newton's *De analysi*, and he did. Leibniz looks through the Newton document and takes notes. Keep that in mind; that will figure in the great war that's coming.

Leibniz goes back and he knows this fellow Newton is really good, and so Leibniz writes a letter to Newton introducing himself, asking about his mathematics. Newton responds with a letter that is now called the *Epistola prior*, the earlier letter. Then Leibniz responds to that and Newton sends a second letter, the *Epistola posterior*, at which point Newton says I've had it, I don't need to write letters to people and he quits the correspondence. But, in this limited correspondence, you can see the two great geniuses feeling each other out, sort of what have you got, what have you got, and they would share little bits of their mathematics, give little hints at calculus, but not reveal the whole picture. They were both playing the cards close to the vest.

In 1676, Leibniz is called back to Germany. Along the way, he goes up to the Hague and meets Spinoza. These two great philosophers spend an interesting afternoon of philosophical chitchat. Then Leibniz consolidates his ideas on calculus, it takes awhile, but in 1684 he publishes a paper on calculus. This is the first paper ever published on the subject; it appeared in a journal called the *Acta Eruditorum* and here it is. This is the first page ever seen in print about this subject of calculus. It's from Leibniz 1684. It's called *A Nova Methodus*, a new method he's going to give you for some really great mathematics.

The title, which is in Latin, of course, translates this way: "A new method for maxima and minima and also tangents, which is impeded neither by fractional nor irrational quantities and a remarkable type of calculus for this." For Leibniz calculus was a set of rules, this method that he was developing that would allow you to do these great problems. If you look at the title, it includes the words maxima and minima, so he's doing maximum, minimum problems and also tangents. Tangent problems—if you know calculus, you know what he's doing here with maximum tangents. That's differential calculus and indeed that's what this paper was, Leibniz's introduction to differential calculus.

The paper itself I should say was not very easy to read. It was pretty opaque. He must've been taking guidance from Descartes. He wrote it in a way that it was very hard to follow. As we'll see a few of his disciples named the Bernoullis had to come along later and try to make this understandable to mere mortals like the rest of us, but at least it was, the rules were there, the calculus was there, and the notation was there. Leibniz introduces the symbol dx to be the differential of x, by which he meant the infinitely small amount by which x grows. If you have a value x and you increase it by a teeny-weeny infinitesimal amount, that little extra piece, that differential, is called dx. The notation has stuck. When we write calculus today, we use dy's and dx's, Leibniz's notation.

He was very interested in getting good notation. He believed that if the notation was well chosen the notation could sort of carry the problem along. You would just push the symbols along and the problem would almost solve itself and so he put a lot of time and thought into his notation. By contrast, Newton never thought the notation mattered very much; he thought it was the substance that was important. His notation was inferior and we don't use it today so Leibniz won the battle of the notation.

In this paper, the 1684 paper, by the end of the second paragraph, not very deep into the paper, he's already given you all the rules of calculus you'll ever want to see. He's shown that if a is a constant, the differential of a is 0. That is in fact true. If a quantity is constant there is no little change in the value of it, d of a is 0. He showed that the differential of ax is adx, the constant a can come out. He stated that the differential of the sum and difference is the sum and the difference of the differentials in this expression, $d(z - y + vv$, notice that instead of v^2, $vv + x)$ is $dz - dy + d(vv) + dx$. There's the differential rule for sums and differences and he gave us the product rule for differentials, $d(xv)$, the differential of the product, is $xdv + vdx$, that's still called the product rule in calculus.

Actually Leibniz initially thought that the differential of the product was the product of the differentials. He thought the d of xv should be $dx \times dv$, which isn't right, but a lot of students today still make that mistake. So did he initially, but he was a smart guy and he figured out the right rule, which is this one. He figured out the rule for quotients, the differential of v/y he said is $\pm\ vdy$ – or $+\ ydv/yy$ where we of course would write y^2 in the bottom.

These are the rules of differentials. If you take a modern calculus course you spend weeks trying to get through these. He just put them all in the second paragraph and moved on. He gave no proofs for these. He just said here they are, use these rules, and they will help you in my new method to solve wonderful problems. Needless to say, a whole lot of work was going to have to be done after this paper to clean things up, but there it is, it's in print, it exists.

Two years later in 1686, he publishes the first paper on integral calculus. The inverse process of taking differentials is doing integrals; here it comes in Leibniz's 1686 paper in which he includes the notation of the integral sign, this squiggly thing, which he thought of as an elongated "S" for summation. He's imagining that he's summing infinitely many infinitesimals and to indicate that he wrote this long S, this beautiful symbol, the integral sign, which now more than anything says that higher mathematics is being done. If you see that sign, there's higher math around somewhere. Laplace wrote that Leibniz had a "very happy notation" and you know that notation and other symbols that Leibniz introduced have stuck with us, very happy.

From the years 1685–1716, Leibniz is pulled out of his mathematical work and other work and sent on a mission to discover, write about, and collect the genealogy of the House of Brunswick, his employer in Germany. This monarch wanted to know where he came from, what were his family roots, and Leibniz your job is to write a genealogy. This became Leibniz's focus. He had to dig through records in various churches, various municipalities; he had to go to Italy to do this. Year after year after year he devoted to this rather meager or minor task of collecting genealogical records of some small monarch in Germany. He could've been doing more math, more philosophy.

It reminds me of the years Newton spent on his alchemy and his theology, which essentially were wasted at the time. We would like to have this time back and let Leibniz do some more valuable things with it. He never finished. He died and he had thousands of pages of notes and all, but never delivered the goods. He did however during this time have the chance to found the

Berlin Academy. He had seen the Royal Society, this group of scholars in London, there was the Paris Academy; he thought Berlin should have one and so he set about founding this. This would play a big role in mathematics in the 18th century. We'll see this Berlin Academy again; Leibniz was its creator.

Alas for the last years of his life he got involved in the priority dispute with Newton. This is the battle between Newton and the British versus Leibniz and the Europeans over who created calculus, who deserves credit for this wonderful subject. This will be the focus of my next lecture, the calculus wars between Newton and Leibniz, but it consumed much of his time and unfortunately was not the most glorious moment for either Newton or Leibniz, as they went to war over the calculus origins.

I should mention that in 1714, Georg Ludwig who is the patron of Leibniz who was the monarch back in Germany, strangely enough was tapped to become King George I of England. I guess what had happened was Queen Anne had died in England without heirs. The British needed some monarch and they checked the genealogy, and sure enough it was Georg Ludwig over here in Germany that should come and become King George I. This would've been a wonderful opportunity for George to bring with him his employee Leibniz and then Leibniz would've been in London exactly when Newton was in London and Newton was running the Royal Society, working in the mint. The two great geniuses finally could've come face to face, and if they had a priority dispute over calculus, they could've duked it out right there in the streets of London—but it didn't happen that way. Georg Ludwig said I'm not bringing you Leibniz; you're not done yet with that genealogy, get to work. Poor Leibniz didn't get to go with the king to England. He had to stay behind with this thankless, endless task, and in fact he died in Germany in 1716.

Unlike Newton who was buried in Westminster Abbey, buried like a king, remember, Voltaire said, Leibniz's funeral was a very small affair. It was said he was buried more like a common thief than what he was, which was an icon who should've been celebrated from one end of Germany to the other. It was a sort of sad and unfortunate ending.

I'll leave you with an image here of the Leibniz statue at the University of Leipzig, his alma mater. You'll see, I think, a very striking comparison here. On the one hand you have Leibniz dressed in the garb of the late 17th, early 18th century, the wig and the buckled shoes and the stockings. He looks very much out of the past. Beside him is the present, the modern future that came after. You look at this and you can't help but think that that future was due in large part to Gottfried Wilhelm Leibniz.

The Bernoullis and the Calculus Wars
Lecture 17

> The Bernoullis were notoriously argumentative, combative, cantankerous, and contentious. They wanted always to have the glory aimed at them; they resented it when the glory was going elsewhere. I like to say that they were the kind of people that gave arrogance a bad name.

Neither Isaac Newton nor Gottfried Wilhelm Leibniz ever married or had children, but Leibniz had the next best thing in his two dedicated disciples, the Bernoulli brothers of Switzerland. Jakob Bernoulli, the older of the two, was a very accomplished mathematician. He investigated infinite series and published, posthumously in 1713, the *Ars Conjectandi*, *The Art of Conjecturing*, a treatise on probability theory. In particular, this work proved one of the foundational results of probability, the law of large numbers. With Leibniz and his younger brother Johann, Jakob also helped refine the subject of calculus.

Johann Bernoulli mentored a young lad named Leonhard Euler, the subject of our next lecture, and was hired to provide calculus lectures to the Marquis de l'Hospital, a French nobleman. L'Hospital later appropriated the written work Johann had sent him and published the first-ever calculus textbook in 1696, although he acknowledged that the work belonged to Leibniz and the Bernoulli brothers. Interestingly, the book covers similar topics as those seen in introductory calculus texts today. A max/min problem posed by l'Hospital is as follows: "Among all cones that can be inscribed in a sphere, determine that which has the greatest convex surface." Just as students do today, l'Hospital solved the problem by taking differentials and setting them equal to zero. The answer is that the height of the cone should be 2/3 the diameter of the sphere.

As you recall from the last lecture, the calculus wars developed between the British and Newton and the Europeans and Leibniz over who should get credit for this innovation in mathematics. In 1708, the Royal Society, headed by Newton, attributed the invention of "the Arithmetic of fluxions" to Newton himself, subtly levying a charge of plagiarism against Leibniz. Leibniz, of course, was infuriated and demanded an investigation to clear his name. Near the end of 1712, the Royal Society—again, headed by Newton—concluded that Leibniz's knowledge of calculus had come after his visit to London and his brief correspondence with Newton in 1676. Leibniz countered that Newton had known nothing of calculus until reading Leibniz's paper of 1684. The battle between the two giants raged on until the death of Leibniz in 1716. It's now clear that they both discovered calculus independently.

The most famous battle in the calculus wars involved Johann Bernoulli, who came to be known as "Leibniz's Bulldog." In 1696, Johann issued a challenge to solve the brachistochrone problem: Assign to a mobile particle M the path AMB, along which, descending under its own weight, the particle passes from A to B in the briefest time. The answer, Johann noted, is not a straight path. By 1697, when no one except Leibniz had solved the problem, Johann mailed it directly to Newton, who quickly found a solution. In fact, the curve of quickest descent is the cycloid curve.

> In 1696, Johann issued a challenge to solve the brachistochrone problem: ... By 1697, when no one except Leibniz had solved the problem, Johann mailed it directly to Newton, who quickly found a solution.

We might, perhaps, settle the calculus wars with a quotation from the mathematician Wolfgang Bolyai: "... it seems to be true that many things have, as it were, an epoch in which they are discovered in several places simultaneously, ... just as violets appear on all sides in the springtime." So it was with calculus. ■

Suggested Reading

Boyer, *The Concepts of the Calculus*.

Dunham, *The Calculus Gallery*.

Edwards, *The Historical Development of the Calculus*.

Grattan-Guinness, *From the Calculus to Set Theory*.

Hald, *A History of Probability and Statistics and Their Applications before 1750*.

Hall, *Philosophers at War*.

Newton, *The Correspondence of Isaac Newton*.

Tent, *Leonhard Euler and the Bernoullis*.

Questions to Consider

1. Read about the wider Bernoulli clan. Besides the original brothers Jakob and Johann, other family members left their marks in mathematics and physics. Chief among these was Daniel Bernoulli (Johann's son), who gave us Bernoulli's principle in fluid dynamics, the principle behind the flight of airplanes and so much more.

2. An infinite series that especially intrigued the Bernoulli brothers was the so-called harmonic series: $1+\frac{1}{2}+\frac{1}{3}+\frac{1}{4}+\frac{1}{5}+\cdots$. Here's a clever argument by which Johann proved the series (in his words) "is infinite," which means its sum grows without bound: For the sake of contradiction, Johann assumed the series had a *finite* value, say H. He then grouped the series into pairs of terms, as follows: $H=1+\frac{1}{2}+\frac{1}{3}+\frac{1}{4}+\frac{1}{5}+\frac{1}{6}+\cdots=\left(1+\frac{1}{2}\right)+\left(\frac{1}{3}+\frac{1}{4}\right)+\left(\frac{1}{5}+\frac{1}{6}\right)+\cdots$ Within each pair, the first number is always larger than the second, i.e.: $1>\frac{1}{2},\frac{1}{3}>\frac{1}{4}$, and so on. Thus, we get: $H=\left(1+\frac{1}{2}\right)+\left(\frac{1}{3}+\frac{1}{4}\right)+\left(\frac{1}{5}+\frac{1}{6}\right)+\cdots>\left(\frac{1}{2}+\frac{1}{2}\right)+\left(\frac{1}{4}+\frac{1}{4}\right)+\left(\frac{1}{6}+\frac{1}{6}\right)+\cdots$ from which it follows that $H > H$ (why?). This is obviously impossible for any finite quantity; thus, by contradiction, Johann concluded that the sum of the harmonic series cannot be a finite quantity.

3. There are many other proofs that the harmonic series diverges to infinity. You might want to find a few of these for the sake of variety.

The Bernoullis and the Calculus Wars
Lecture 17—Transcript

Neither Isaac Newton nor Gottfried Wilhelm Leibniz ever married. They had no children to pass along the calculus gene, but Leibniz had the next best thing. He had two dedicated disciples, the Bernoulli brothers of Switzerland. These are the Bernoulli's; the older brother is Jakob Bernoulli, the younger brother is Johann. Perhaps you can see some resemblance there; I don't know, maybe in the sneer that they're giving. The Bernoulli's were known as rather hard, sharp people.

I want to introduce them to you, talk about their mathematics, and talk about Johann's role in the calculus wars between Newton and Leibniz. Let's first meet Jakob, the older brother. He was a very accomplished mathematician. He investigated infinite series, this subject that keeps coming up as being so important at this time. He also published, posthumously in 1713, the *Ars Conjectandi*, The Art of Conjecturing. This was the greatest treatise to date on probability theory.

We've seen that Cardano had written a book about probability as it related to gambling so he could win more at the table. Then we saw that Fermat and Pascal had exchanged letters that helped establish the study of probability. But, this was a real beautiful tome, the *Ars Conjectandi*, that pushed the boundaries of probability theory even further. In particular, Jakob, in this work, proved the law of large numbers.

If you know probability, you know this is one of the foundational results in the theory of probability. It says, more or less, that if you do an experiment over and over again, a large number of times, the proportion of the times you'll see a success will get very close to the true probability of that success. If you flip a coin a thousand times, you're going to get pretty close to 50 percent heads and 50 percent tails assuming the coin is balanced. But, if you flip it a million times you're going to get even closer to 50 percent heads and 50 percent tails. The large number of trials starts to yield this probability. It's a critical result. It's a theorem and Jakob proved it. He was very proud of that.

He also helped push forward the frontiers of calculus. In fact, with his brother Johann, Jakob and Leibniz formed the kind of triangle who help refine the subject. I had mentioned that Leibniz first published the calculus in a paper that was, the Bernoulli's called it, more an enigma than an explanation. It was hard to follow. Jakob and Johann Bernoulli would talk with Leibniz and they'd suggest maybe we should change this or why don't we try it this way. The three of them together refined the subject into what we know it as today.

One other thing—Jakob suggested the name "integral" calculus. Leibniz had talked about calculus differentialis and calculus sumitoris, the sumitori and the differential calculus, and Jakob said no. It doesn't sound good, that sumitoris thing; how about integralis calculus? Integralis, oh yeah that's the name, and that's the name that stuck. Jakob had quite a legacy. So did younger brother Johann. What did he do? He too investigated infinite series, made good progress, and raised a very important question or two about these incredible mathematical entities. He too pushed the frontiers of calculus and he did two other things of particular interest. One is he mentored a young lad named Leonhard Euler who happened to live in Johann's hometown of Basel, Switzerland and young Euler would grow to become the dominant mathematician of the 18th century in the subject of our next lecture. Johann was his teacher.

Johann Bernoulli was hired to provide calculus lectures to the Marquis de l'Hospital. This is a kind of curious story from the history of mathematics. This is the multiply named Guillaume François Antoine de l'Hospital, a French nobleman. I think you've got to agree this is the greatest wig we've seen yet. This is a nobleman's wig. L'Hospital wanted to learn about this newfangled subject called calculus he'd been hearing about, but he needed help. In 1691 Bernoulli, Johann Bernoulli, met l'Hospital in Paris and l'Hospital said let me hire you to teach me this new subject.

By 1692 Bernoulli is lecturing l'Hospital on calculus either physically lecturing him, speaking to him, or sending him documents he's written. Bernoulli would write something, mail it to l'Hospital and then that way teach him the subject. The two had an agreement, a monetary agreement. Here was l'Hospital's stipulation. He said to Bernoulli, "I shall give you with pleasure a pension of three hundred livres … I ask you to give me occasionally some hours of your time to work on what I shall as you – and also to communicate to me your discoveries, with the request not to mention them to others."

l'Hospital said you know you send me your works, I will send you money. l'Hospital was the rich nobleman, Bernoulli was the professor, not very rich. Just don't tell anyone what you're doing says l'Hospital; just send it to me. l'Hospital takes all of this that he learns from Bernoulli and a lot of the written work that Bernoulli has sent him and publishes in 1696 the first calculus textbook ever. This is it, *The Analysis of the Infinitely Small for the Understanding of Curved Lines*—this is the title.

What this is, is a compendium of Bernoulli's writings. Johann Bernoulli essentially had written this, but l'Hospital compiled it, edited a bit, and got it published and so now there is a calculus textbook. Leibniz has published the papers on calculus in the 1680s, but now there's an actual book in 1696. In the preface l'Hospital writes this; he's referring here to the Bernoulli brothers and to Leibniz. He says: "I have made free use of their discoveries, and I frankly return to them whatever they please to claim as their own." He's not asserting that he made all this up. He acknowledged that he was borrowing things from Leibniz and the Bernoulli's to compile this calculus and if they want the rights to it that they may claim them.

Here's the title page, the table of contents for the first calculus book ever. Section 1 was On the Rules of Differentials and that's where we see these rules on what's the differential of the sum, what's the differential of the product, the kinds of things that appeared in Leibniz's original paper although here l'Hospital proves them, gives reasons why they should work. Now you've got the rules. In Section 2, he uses Differential Calculus for Finding Tangent lines, one of the great things you do with calculus.

The next section he shows you how to use Differential Calculus for Finding Greatest and Least Values, hmm that sounds like max/min problems. Section 4, Using Differential Calculus for Finding Points of Inflection, if you've taken differential calculus, usually Calculus I as it's called nowadays, chances are this is your table of contents. It hasn't changed much over the centuries. We're really quite fortunate that this first book on calculus was so good, l'Hospital's book really was a masterpiece and it got the subject off to a good start.

Let me mention a result in there just to give you the flavor of this first calculus book. This is Example 6 from Section 3. l'Hospital poses this problem, "Among all cones that can be inscribed in a sphere, determine that which has the greatest convex surface," greatest surface a maximum. Here we are in a max/min problem and what he is looking for is the greatest surface of the cone, the convex surface would be the slanty part, not the bottom, not the circle on the bottom, but the sides.

Let me show you a picture here. Suppose you have a sphere, when I inscribe within it a cone, what that means is let's say we'll put the vertex of the cone up at the top and then the circular base of the cone will be a circle on the sphere, but you can imagine putting it in various spots. I mean you could put it way up near the top for instance. There's a picture of the cone inscribed up near the top of the sphere. Its surface wouldn't be all that great. It's not that big a cone. You could lower that circle of the cone, maybe put it at the equator and now get a cone with the same vertex up at the top it would probably have more surface. You could even put the circle down below the equator such as in this picture where I have the sphere and my cone now sort of dips below the equator or you could actually put the cone so that the vertex is at the top and the circle where it meets the sphere way down at the bottom. That would be a long, skinny cone, probably not much surface.

What l'Hospital's saying is as you move this cone around up and down the sphere the surface will grow and then shrink, how big can it be? Where do you put this cone to get the greatest convex surface? It is not a trivial question. What's he do? He sets up an equation. He takes the differentials; he sets them equal to zero. He solves; he gets the answer. It's just how we do it today in differential calculus. In fact, this is a great problem even today in differential calculus. I like to give it to my students and see if they can do a problem from the first calculus book. What's the answer? The answer is this: The height of the cone should be 2/3 the diameter of the sphere. This is what the calculus yields. You make your cone go down 2/3 the diameter of the sphere. That will give you the maximum surface area. It's a wonderful problem from the first calculus book.

I should mention another result from that l'Hospital calculus book in Section 9, there appears what we now call l'Hospital's rule, which is a famous result in differential calculus. We call it l'Hospital's Rule because it's in l'Hospital's calculus book, but it actually wasn't his rule. He didn't invent it, it was made up; it was discovered by Johann Bernoulli who passed it along dutifully in exchange for the money that was coming the other way. Bernoulli discovers this great rule, sends it to l'Hospital, l'Hospital publishes it, and it gets called l'Hospital's rule. Ever since, really we should call this Bernoulli's Rule shouldn't we? We want to give him credit, he discovered it, but the 20th-century math historian Dirk Struik said no, no. Said Struik, "Let the good Marquis keep his elegant rule; he paid for it." He bought the rights and so maybe we'll just continue to call it l'Hospital's Rule.

The Bernoullis were notoriously argumentative, combative, cantankerous, and contentious. They wanted always to have the glory aimed at them; they resented it when the glory was going elsewhere. I like to say that they were the kind of people that gave arrogance a bad name. If you want to be alliterative you can call them the "Brilliant but bickering Bernoulli brothers." They would've been really difficult people to know. Although they liked to argue with others they particularly liked to argue with each other. They didn't get along very well much of the time. The older brother Jakob had cornered the market in the Bernoulli family on mathematical expertise and along comes younger brother Johann who's just as good. That undoubtedly called some sibling rivalry, which they never really got over.

Let me tell you a story that illustrates the "brilliant but bickering Bernoulli brothers." The year is 1691 and the problem involves the challenge of the catenary curve. This requires a little bit of background. What is that? Suppose you took two nails and pounded them into the wall and hung a chain from these two nails. The chain would hang down in a shape something like this. That shape, the shape of a hanging chain is called the catenary. It actually comes from the word in Latin, the word catenary means chain, it comes from the Latin root for that. The issue is what's the equation of that? Can you use your mathematics to figure out the equation of the hanging chain, the equation of the catenary? Jakob Bernoulli, the older brother, wanted to find this out. He had worked on it for the better part of a year and it stumped him.

He couldn't do it. It's not simple. It kind of looks like a parabola if you look at it, but it isn't. It's way more sophisticated and he was stuck.

What does Jakob do? He publishes his problem and he says I can't do it. Can anybody do this problem? This goes out to the wider audience. Can anybody help me figure out the equation of the hanging chain? Jakob got stuck; he couldn't do it, but younger brother Johann looks at it and gets an idea and solves it. Of course this was good news for younger brother Johann and bad news for older brother Jakob.

Let me tell you how Johann described this. This will be Johann speaking, the younger brother who could solve it and figured out what the catenary was as he reports what happened next. Johann says, "The efforts of my brother," that is Jakob, "were without success; but for my part," says Johann, "I was more fortunate, for I found the skill and I say it without boasting, but why should I conceal the truth? I found the skill to solve it in full. It is true that it cost me study that robbed me of rest for… an entire night …" Yikes, an entire night? His brother had been working on this for a year. Johann says he then ran to his brother and showed him the answer and told his brother to stop working on it. I've solved it; I Johann have solved it. After this they hardly spoke to each other. This was not a good moment. But that's the Bernoullis.

Let me get to the Calculus Wars. The Calculus Wars remember developed between the British and Newton and the Europeans and Leibniz, who should get credit for calculus. Newton read Leibniz's papers in 1684 and '86, Newton sort of thought maybe Newton's name should appear in there somewhere, but it didn't and Newton was a little grumpy right from the get go. But things heated up in 1708 when the *Philosophical Transactions* of the Royal Society of London described, "… the Arithmetic of fluxions which Mr. Newton, beyond all doubt, first invented; the same Arithmetic, under a different name and using a different notation, was later published, however, by Mr. Leibniz." This appears in the Royal Society's transactions. If you read that it isn't an explicit charge of plagiarism, but it's close. Right, it says Newton discovered this and Leibniz sort of changed the name, changed the notation, and later published however. It's rather dismissive. It seems to be pretty clear who the Royal Society thought deserved credit. You might want to remember who was the head of the Royal Society in 1708—Newton. Who wrote this? Newton did, so he takes a stand.

Leibniz reads this and he's infuriated. He thinks he's been charged with stealing the subject and so he demands an investigation. He wants his name to be cleared. He asked the Royal Society to look into this. That's a big mistake because who's running the Royal Society, who's going to be his judge? Newton is judge. This didn't work for Leibniz and near the end of 1712 the Royal Society issues its verdict. It's called the *Commercium epistolicum* and it asserted, it concluded that Leibniz had had no inkling of calculus until after 1677.

You've got to stop and remember some dates here. But remember Leibniz had gone to England in 1673 and in 1676 had seen Newton *De analysi*, taken some notes on this document. Then there had been those letters exchanged, the *Epistola prior*, the *Epistola posterior*. All of that happened before 1677. What the Royal Society said was only then did Leibniz have any idea what calculus was; conclusion he stole it. He gathered the information from Newton's manuscript and from these letters and only then did he have the subject in hand, which he then published before Newton had a chance. That was sort of the conclusion.

Leibniz was guilty according to the Royal Society. Leibniz wasn't going to take this. He responds in 1713, "It appears that there is room to doubt whether he knew my invention before he had it from me." He's responding to Newton, maybe Newton didn't know calculus until he read it in my paper. Leibniz after all had published first, how do we know that Newton had it in his desk. Maybe he just learned it from Leibniz. So these two giant figures are going at each other pretty seriously. Leibniz said in 1714 that "… the new discoveries that were made by the help of differential calculus," which remember was what Leibniz called it, "were hidden from the followers of Newton's method, nor could they produce anything of value …" anything of real value.

Leibniz was certainly rooting for himself on this one. The battle continued, the war raged on until Leibniz died in 1716 and kind of lost by default I guess. No one else was going to be arguing for him. Those are the wars. What's the conclusion? What happened? Who does deserve credit for this? I mean calculus is arguably the greatest discovery in the history of mathematics, how do we resolve this? The answer is the following. They both deserve credit. They both discovered it independently. This is now clear. It's true that Newton got there first; he had worked on it first while he was a student at Cambridge. Leibniz, however, had possession of all the rules before he went to London, before he wrote the letters and exchanged the correspondence with Newton, before he saw the manuscript from Newton. We have Leibniz's notes from 1673 and 1674 and it's all there, even the integral sign, the dx, the differential symbol. Leibniz was on top of this. It's true he saw Newton's manuscript and it's true he took notes, but the notes weren't about anything that looked like fluxions or calculus. The notes were about some of the neat infinite series work that Newton had done. That Leibniz thought was new and wonderful, but the calculus stuff Leibniz already had. We're not going to have to decide; we're going to give them both credit.

Here are some comparisons. If you said who did it, I'm saying Newton did it and he called it fluxions, I'm saying Leibniz did it and he called it calculus. Where was it done? Newton did it in Cambridge and Leibniz discovered calculus in Paris. If you want to do a vacation tour to the origin of calculus there's where you want to go, Cambridge and Paris, it would be a great vacation. When was it done? Newton comes first, 1664–1667, his *Anni Mirabiles*, Leibniz does it during his wonderful years as a diplomat in Paris a decade later. But, when was it published? Newton published a little piece of his fluxions in 1704, but as we mentioned the full-blown version of his fluxions wasn't published until after his death whereas Leibniz beat him to the punch with the 1684 and 1686 papers. If you want to count priority in terms of who did it first chronologically Newton. Who did it first in print? Leibniz did. I'm going to say they both get credit.

What remains then is the most famous battle in the calculus wars and it involves Johann Bernoulli who came to be known as "Leibniz's Bulldog." He was Leibniz's disciple. He revered Leibniz and he was going to go to bat for his hero Leibniz against those British that he didn't like very much. He figures prominently in this story of the challenge problem of the Brachistochrone. Here's the problem. The year was 1696, Johann Bernoulli had solved this problem and now he writes it up and puts it out as a challenge to the world. He says here's a new problem, which mathematicians are invited to solve and here it is. Here's the problem. Assign to a mobile particle M the path AMB along which, descending under its own weight, the particle passes from A to B in the briefest time. Brachistochrone, chromos means time, brachisto means shortest. We want to get from A to B rolling a marble say down a ramp of some shape fastest. What's the one that'll get you from A to B in the briefest time? That curve is called the brachistochrone; that's the curve that Johann Bernoulli had brilliantly been able to discover. He knew what it was and now he's challenging the world to see if other people can send him their solutions. See who else is as good as he is.

Johann writes this, this is still in the challenge in the statement of the problem. He says, "To forestall hasty judgment, although the straight line AB is indeed the *shortest* between the points, it nevertheless is not the path traversed in the briefest time." Hang on here. He says you know before you send me the straight line that's the wrong answer. Initially you'd think that is the right answer isn't it, from A to B you'd sort of go down the straight line because that's the shortest distance. But, if you think about it, if I let a marble go and it's rolling down a straight line, it's going to go kind of slowly at first and it will pick up speed. It doesn't have so far to go, but it doesn't start off very fast.

Suppose you had some sort of arc instead and you let the marble go, it would fall almost vertically at the outset and pick up a lot of speed. If you did this just right it might actually go so fast that although it has a further distance to travel it might beat the marble that's rolling down the ramp to the finish line and in fact Bernoulli says the answer if not the straight line. "However," he writes, "the curve AMB, whose name I shall give if no one else has discovered it before the end of this year (1696) is one well known to geometers." He's giving a hint. The answer is it's a well-known curve; it's not a straight line. Send me your answers by the end of the year. That's the challenge of the brachistochrone.

What happens when the end of the year comes? He's only got one solution and it's from Leibniz, his hero. But, Leibniz says to Johann Bernoulli, I think people didn't see your problem. They might not have read that issue of the journal where you wrote it. I think you should give people more time. You've got to sort of extend this. Students always like this one. It gives them some more time, extend the deadline. At Leibniz's suggestion Bernoulli restates the problem in the next issue of the journal and extends the deadline. The problem is still there, what's the curve of quickest descent. Now Johann says, "If Easter of 1697 passes and no one is discovered who has solved our problem, then we shall withhold our solution from the world no longer." You've got until Easter and that's it. That's published along with the problem in this second go-round.

Then, Johann couldn't help writing more. This is still in print and he says this, "… so few have solved our extraordinary problem, even among those who boast that through special methods… they have not only penetrated the deepest secrets of geometry but also extended its boundaries in marvelous fashion, although their golden theorems, which they imagine known to no one, have been published by others long before." Who's he talking about? Who had his results published by others long before, but claimed that they were his golden theorems? That's definitely a slam against Newton. This is in print; Bernoulli attacks Newton right there, who hadn't submitted a solution. Then just to be sure Newton didn't miss this, Johann Bernoulli takes an issue of this journal with the challenge in it, puts it in an envelope, and mails it to Newton in London. Newton is not going to escape the challenge.

Let's hop the Channel and pick up the story in England. At this time, Newton was at the mint working hard. He was living with his niece and she report this, she said, "When the problem in 1697 was sent by Bernoulli, Sir Isaac was in the midst of the hurry of the great recoinage [and] did not come home till four from the Tower very much tired… but did not sleep till he had solved it which was by four in the morning." The old man, Newton, trudges home after a hard day at the office, finds the challenge, reads this attack upon him, and solves it by four in the morning. It has been said that this showed Newton was getting old; that the young Newton would've solved it by midnight. In any case, he gets the answer, he writes the solution down, doesn't sign it, and sends it anonymously back to Bernoulli.

What is the answer here? The answer what is the brachistochrone? What is the curve of quickest descent? It's the cycloid curve. That's the curve it is. We've seen this before. Pascal had studied the cycloid. It was this amazing curve; remember what it was? It was if you rolled a circle along a horizontal line and followed a single point on the circle, it would trace out these cycloidal arches except what you do is you'd drop this below the curves so now the cycloidal arches are going this way and if you roll down a cycloid from the point A to the point B that's the curve of quickest descent. It's an amazing problem.

When Easter arrived, five answers had come in to Bernoulli. Here's who could do it. Johann Bernoulli could do it, of course he could; he only made the challenge because he knew the answer. He was one. The "celebrated Leibniz" sent in a solution, his hero, two. Jakob Bernoulli sent in an answer, his brother. I'm sure Johann was upset by this, but his brother did it, so that's three. l'Hospital sent in a solution. Makes you wonder if he bought it from somebody, but in any even l'Hospital got one, and then there was this anonymous solution that arrived from London. Johann Bernoulli was not a particularly gracious person, but he knew he'd been beaten. He knew where this had come from and famously he said, "I recognize the lion by his paw." I recognize Newton; he's the solver. He's my anonymous source and so Bernoulli concedes that he's been beaten in the brachistochrone skirmish. Newton was not quite so gracious. Newton's response was, "I do not like to be teased by foreigners about mathematical things."

Let me end with the two great discoverers of calculus one last time. There they are, Newton and Leibniz, looking down upon us and let me give you a quotation from a later mathematician named Wolfgang Bolyai. He was talking about something else, but it fits very well our story. Bolyai said, "… it seems to be true that many things have, as it were, an epoch in which they are discovered in several places simultaneously, … just as violets appear on all sides in the springtime." So it was with calculus.

Euler, the Master
Lecture 18

> [His hardships make] Euler, really, the counterpart of Beethoven. Remember Beethoven loses his hearing and yet continues to produce great music. Euler loses his vision but continues to produce great mathematics.

Leonhard Euler, born in Basel, Switzerland, in 1707, was history's most prolific mathematician. His genius was evident from an early age. His father wanted him to become a pastor, but Euler's skills and talents seemed better suited to math and science. Thus, in 1720, when he was 13, it was arranged that he would study with Johann Bernoulli. In not too much time, Euler would pass his master, and Bernoulli would stand in awe of what his former student could do.

In 1722, at the age of 15, Euler graduated from the University of Basel; he then completed a master's degree. In 1727, at age 20, he took an appointment at the St. Petersburg Academy in Russia, then moved to the Berlin Academy in 1741. In 1766, he returned to St. Petersburg and remained there until his death in 1783.

By all accounts, Euler had a phenomenal memory, which served him well after the 1730s, when he lost vision in one of his eyes, probably the result of an infection. In 1771, he lost vision in the other eye as a result of botched cataract surgery. Essentially blind, Euler nonetheless continued his active career. In 1775, for example, he produced 50 papers, dictating them to a group of young scribes. The collected works of his lifetime run to more than six dozen volumes—25,000 pages of published mathematics. The first volume of Euler's work was published in 1911, and the project to complete publication of his work is still ongoing a century later. The quality of his output is equally as astounding as the quantity. Three

Three of Euler's theorems were voted by mathematicians into the top five most beautiful theorems of all time, and almost 100 mathematical terms carry his name.

of Euler's theorems were voted by mathematicians into the top five most beautiful theorems of all time, and almost 100 mathematical terms carry his name. Further, Euler wrote great textbooks, including the *Introductio in analysin infinitorum* (1748), which introduced functions as the critical entity of mathematical study. He also wrote texts for differential and integral calculus, as well as books on mechanics, optics, and popular science. His bestselling book ever was *Letters to a German Princess*, a series of essays on elementary science.

Highlights of Euler's achievements include the number e, which denotes the base of the natural, or hyperbolic, logarithm. This number still plays a critical role in calculus. He also gave us Euler's identity, $e^{i\pi} + 1 = 0$, an expression that contains what are, perhaps, the five greatest numbers. The Euler polyhedral formula, V

Euler's blindness when he was in his sixties did not interfere with his prolific career.

(vertices) + F (faces) = E (edges) + 2, is another famous result, pervasive in the realm of solids. Still another great result is the solution to Basel problem; here, the task is to find the exact sum of the following infinite series: $1 + 1/4 + 1/9 + 1/16 + 1/25$ …. In the next lecture, we'll see Euler's solution, which is exactly $\pi^2 / 6$. What's called the Euler path was suggested by the question of whether the citizens of Königsberg could travel a route around their city in which they crossed each of its bridges once and only once. Euler's solution launched the discipline of graph theory. In geometry, he found something called the Euler line of a triangle, on which the orthocenter, the centroid, and the circumcenter align.

Condorcet, the great French mathematician, had this to say about Euler: "All celebrated mathematicians now alive are his disciples: there is no one who is not guided and sustained by the genius of Euler." ∎

Suggested Reading

Biggs, et al., *Graph Theory: 1736–1936.*

Dunham, *Euler.*

Euler, *Elements of Algebra.*

———, *Introduction to Analysis of the Infinite.*

Fellman, *Leonhard Euler.*

Heyne and Heyne, *Leonhard Euler.*

Maor, *e: The Story of a Number.*

Richeson, *Euler's Gem.*

Sandifer, *How Euler Did It.*

Tent, *Leonhard Euler and the Bernoullis.*

Questions to Consider

1. Check the Euler polyhedral formula ($V + F = E + 2$) for polyhedra in the shape of the Great Pyramid of Cheops, the U.N. Building, and the Pentagon. Recall that V is the number of vertices (corners), F is the number of faces, and E is the number of edges in the polyhedron.

2. If you like geometrical constructions, get out a large piece of paper, draw a big triangle on it, and proceed to construct the orthocenter (where the three altitudes meet), the centroid (where the three medians meet), and the circumcenter (where the three perpendicular bisectors meet). If you do this carefully, you should see that these three points line up, with the centroid half as far from the circumcenter as from the orthocenter. This is the Euler line. Isn't it amazing that no one had spotted this before he did in 1767?

Euler, the Master

Lecture 18—Transcript

In this lecture, we'll focus on Leonhard Euler of Switzerland. Euler was history's most prolific mathematician. He was certainly the dominant force in 18th century mathematics and he's a person that everyone should meet. What I want to do in this lecture is give his biography and sketch some of his great achievements. The next two lectures will be great theorems to accompany the great thinker as we look at some of his mathematical achievements. I promise the two lectures that follow this will be just spectacular in terms of what Euler did.

But, first we've got to meet him. This is Euler. When people see this picture, their first thought is, what is he wearing? It's certainly a curious outfit to put on when the portrait painter comes over. It looks like he just got out of the shower actually. But, in fact, this was fairly standard garb for the 18th century. What's going on here is this: the folks back then would wear wigs and shave their heads. Their heads would get cold when they weren't wearing the wig. They would wear this peculiar looking headdress just as a head warmer if they were living in cold climes and Euler did. This was just a head warming device. What's interesting is that, when the portrait was going to be done, he decided to dress informally like this rather than put on the wig. That conforms to his basic personality, which was a kind, unassuming person. This is Euler in informal attire.

He was born in 1707 in Basel, Switzerland. His father was a pastor at this church. I'm told that, if Euler were to return to life, he would recognize it. It still looks pretty much as it did back then. His genius was evident quite early on. His father wanted him to become a pastor and follow in his footsteps, but Euler's skills and talents seemed better suited to math and science. Thus, in 1720, when he was 13, it was arranged that he would study with Johann Bernoulli. We met Bernoulli in an earlier lecture. Bernoulli was in Basel, Switzerland at the time.

Bernoulli was probably the world's greatest active mathematician and he was right there in Euler's hometown. Young Euler would go to see the great Bernoulli. He later remembered this as being kind of terrifying. Here was Bernoulli, this crusty, arrogant, and towering figure. Here comes little Euler with his week's work. Bernoulli would criticize it, make suggestions, and send him off to do another bit of mathematical work. At that stage, Bernoulli's obviously the master, Euler the pupil. In not too much time you begin to see them acting as colleagues and then in not too much time beyond that it's Euler who's the master and Bernoulli is in awe of what his former student could do.

In 1722, at the age of 15 Euler graduated from the University of Basel. He then completed a Masters Degree and applied for a job there. There was actually a position in physics open at the University in his hometown. He wanted it and he failed to get it. The job went to somebody else and it became I think the worst hiring choice in the history of the University of Basel, if not any university, they turned, they passed on Euler in favor of somebody who has now been completely forgotten.

What does Euler do? He needs a job and in 1727, at the age of 20, he takes an appointment at the St. Petersburg Academy in Russia. Remember in Europe at this time the big courts would be surrounded by these Academies, the Royal Society in London, the Paris Academy, and the Berlin Academy. Russia wanted one of these and so the St. Petersburg Academy was being created to bring scientists, writers, musicians, and scholars from around the world to reflect well on the court on the tsar, on the nation.

Daniel Bernoulli was already at St. Petersburg. He was the son of Johann Bernoulli and he knew Euler and he sort of said to Euler come on up here we can get a position for you and Euler does. He goes to St. Petersburg in the cold Russian seacoast, stays there until 1741 when he goes to the Berlin Academy. In those days, scholars would jump from Academy to Academy even as sports players do now from one team to another. He was wanted at Berlin. He went there, this was Berlin under Frederick the Great. At the Academy when Euler was there was Voltaire, Dalenbare, people like that so he was mixing with some great names.

He stayed in Berlin until 1766 when he is called back to St. Petersburg and remains there until 1783 when he died. His lifespan from 1707–1783 is almost exactly the same as Benjamin Franklin's. If you want a comparison on this side of the Atlantic, think Ben Franklin living roughly the same period as Euler. Franklin's life of course was quite different. He traveled the world. He was engaged in politics. He was engaged in science. Euler's was a little more constrained. His whole life was spent from Basel to St. Petersburg to Berlin to St. Petersburg. But his intellectual journeys were every bit as amazing as Franklin's. It would've been neat had they met, but they did not. On the personal side, Euler was married and he and wife Katharina had 13 children. You must remember, however, in the 18th century childhood mortality was a very serious and heartbreaking phenomenon and of these 13 children only 5 survived to adulthood, so there must've been real sorrow in the Euler household.

Euler had by all accounts a phenomenal memory. The stories are that he could memorize anything. He could memorize poems, plays, and speeches. He could memorize tables, table of logarithms, and tables of primes. It was much easier than having to look up a number if you could just recall it from your memory and this he could. This memory would serve him well when personal calamity strikes, namely in the 1730s Euler lost vision in one of his eyes. We think now what happened was an infection got loose somehow and went to his eye and not only cost him his vision, it literally destroyed the eye so that the eye collapsed and was of no use to him.

What did he do then? He sort of gave up. He backed off his work, right, wrong. He kept going. The fact that he had lost vision in one eye did not slow him down one bit and he proceeds with his very active research career until 1771 when he lost vision in the other eye. This was a cataract and nowadays that's something that easily fixed with surgery, but in 1771 it was quite a serious problem. They did try surgery on Euler and you don't even want to think about eye surgery in 1771. It was painful and it failed and it left Euler essentially blind. Apparently he could see vague shapes, but nothing more. He gives up right? Wrong, he keeps going, just as active as he ever was. He did not let the blindness stop him. If you don't believe me let me just note that in 1775 as a blind man he produced 50 papers that year, a paper a week and he couldn't see.

What he would do is this. He would have a table full of scribes, young people with their quill pens and their paper and he would dictate his papers and they would furiously write trying to keep up with him, trying to do this mathematics that he could see only in his mind's eye. But, it just kept spewing out. If a new paper came in from somebody else, from Lagrange or Dalenbare they would read it to him and again with this phenomenal memory of his he could capture it and comment and proceed further.

Euler's story, I think, is the most inspirational in the history of mathematics. Here he is faced with this terrible physical ailment and he does not stop. He proceeds as always; he doesn't let it beat him. This makes Euler really the counterpart of Beethoven. Remember Beethoven loses his hearing and yet continues to produce great music. Euler loses his vision, but continues to produce great mathematics. His work is characterized by its remarkable quantity and its remarkable quality, both. He's off the scale in both of these dimensions. Let me mention the quantity first. His collected works are called the *Opera Omnia*. These are now over 6 dozen volumes and 25,000 pages of published mathematics. What happened was in 1911 the Swiss Academy of Science started to publish Euler's collected works when the first volume came out.

This is a volume. Just let me show you this. It's huge, right. It's a big, thick tome, hundreds of pages, full of mathematics. If this was the product of someone's entire life's work, that would be quite impressive, but this has 76 cousins just like it that have been coming out ever since 1911. It's now a century since the first volume came out and they're still coming. They're not done yet. They're not quite sure how much mathematics Euler will eventually fill, how many volumes this will run to. It's going to be in the vicinity of 80, 90, who knows. We'll be deep in the 21st century and they'll still be coming out.

I like to think that the grandchildren of those original editors of those first volumes are now old and yet Euler is still keeping them busy. There's nothing like it in the history of mathematics. He did more mathematics then anybody. In fact, here's an interesting statistic, after he died he was still publishing. There were papers in line to be published. There were things in his desk. There were things in his cabinet that people found. For decades after his death he was still publishing. His publication count as a dead man is 228 papers. I guarantee you no dead people published 228 papers in mathematics except Euler. He was just phenomenal in his quantity of work.

As to the quality of work, equally great! Back in the 1980s a journal called the *Mathematical Intelligencer* surveyed the mathematical community and asked them for a ranking of the most beautiful theorems of all time. You would check off a postcard and send it in and they would do a poll or survey. What are the most beautiful theorems ever? When the results were done, Euler was responsible for 3 of the 5 most beautiful formulas or theorems ever and 5 of the top 16. He sort of topped the charts. He was just amazing in the quality of his work.

Another way to measure that is this. There're mathematical terms you can find in online math dictionaries. You can put in a term, I put in Euler to see how many mathematical terms of note have his name attached to them. I hit 96 entries that came back with Euler's name attached. There's the Euler line. There's the Euler identity. There's the Euler product sum formula. There are Eulerian graphs and Eulerian functions, 96 terms in mathematics carry his name, and you know they don't get into the dictionary unless they're important. Most mathematicians have no terms named after them. Euler has almost 100. That is a suggestion of the extraordinary quality of his work. You could hardly study a branch of mathematics without running into some major theorem that he gave us.

He also wrote great textbooks. He was one of the great textbook writers of all time. One of them was the *Introductio in analysin infinitorum* of 1748. I mentioned this in an earlier lecture, Carl Boyer had ranked the *Elements* of Euclid as the greatest textbook of classical times, Al-Khwārizmī *Algebra* is the greatest textbook of medieval times and this Euler's *Introductio* as the greatest textbook of modern times.

This is the book in which he introduces functions as the critical entity of mathematical study and gives you the functions of importance, the polynomial functions, the exponential functions, the logarithmic functions, the trigonometric functions, the inverse trigonometric functions. If you've taken calculus or pre-calculus you might recognize these. I know these; these are the functions we study. They've been the functions we've studied ever since Euler's *Introductio*. They're the basic functions in our toolkit in mathematics. That was a very significant work. In the 1750s, he wrote a differential calculus text; it became the standard. In the 1760s, he wrote an integral calculus text that actually covered three fat volumes. It became the standard, 1770 an algebra text. This is when his eyesight was failing him. Nonetheless he wrote this massive treatise on algebra. He also wrote applied math. He had a mechanics book, very influential in the 1730s, an optics book, and he wrote popular science.

His bestselling book ever was *Letters to a German Princess* in which Euler explained to a real German princess to whom he sent these. They were later collected and published, but he was actually sending them to a young girl and explaining science. Why is the sky blue, why is it colder on top of a mountain than at the bottom, that sort of thing. They're a bunch of nice essays. They're still in print and it's not always the case that a great research scientist can come down to the level of popular writing. Euler did this quite well.

Gauss said that, "The study of Euler's works will remain the best school for the various fields of mathematics, and nothing will replace it." Gauss was a very great mathematician who much admired Euler. Laplace it is said in response to his students inquiries of how do I learn math advised them to, "Read Euler, read Euler. He is the master of us all." There are very deep footprints from this 18th-century figure.

In the remainder of this lecture I want to show you some of the things he did just to survey some of his achievements. There are 25,000 pages so we're not going to get them all in there, but just some of the highlights. One of them is the number *e*—if you've taken calculus, you've certainly run into this. If you haven't taken calculus you probably think I should be saying the letter *e*, but actually it's a number and here it is as it appeared in his writings. You can see it right there in the middle of the page.

He says he's going to let the symbol *e* stand for the number 2.71828, etc. He carries it out to, I think, 12 places. He loved calculating things numerically. He says which therefore denotes the base of the natural or hyperbolic logarithm. Sure enough that's what *e* is, it still plays that role, it's a critical number in the calculus. At the bottom of the page there you see he shows you how to calculate it, $1 + 1/1 + 1/1 \times 2 + 1/1 \times 2 \times 3$ and so on, an infinite series. It converges very rapidly to e, you don't have to calculate too many terms and you get this real accurate estimate. That's Euler's number. If he had done nothing, but give us e, we'd remember him to this day.

But, he did so much more. One thing he did in 1748 was give us Euler's identity. This is it in modern notation, e, that number e to the ix power is $\cos x + i \sin x$. Then, i is the square root of -1, the imaginary constant. This is a very weird formula. On the left side we have an exponential, on the right side we have trig functions, sin's and cos's, there's i's flying around here, what a strange wonderful relationship which he proved. Here's how he printed it, $e^{+v\sqrt{-1}}$ he wrote. At that time he hadn't adopted i for the $\sqrt{-1}$, he had to write it out. Then on the right side he says equals cos. $v + \sqrt{-1} \times \sin. v$; this is because cos and sin were abbreviations for cosines and sines and since he was abbreviating he thought he better put the period in there. We now don't use the period.

If you take Euler's identity and let $x = \pi$ in this, you get $e^{i\pi} = \cos \pi + i \sin \pi$ and if you reduce this in terms of your trigonometry values, your trigonometric knowledge it boils down to $e^{i\pi} + 1 = 0$. In that poll I mentioned of the most beautiful theorems of all time, this was number one. This was regarded by the mathematical community as the most beautiful theorem ever, $e^{i\pi} = 1 = 0$. It's such a peculiar relationship and what makes it so intriguing is that it contains the five greatest numbers in the world.

If you were going to have a party and you wanted to invite the five greatest numbers to your party whom would you invite. You'd invite zero. Zero is the additive identity; it's the number that when you add it to something it leaves it unchanged. That's a critical number, zero. You'd invite one. That's the multiplicative identity. When you multiply something by one it's unchanged. That's pretty important. You better invite e if you want to do calculus. You'd better invite π if you want to do geometry and you better invite i if you want to do anything with a complex realm. There they are the five great constants, 0, 1, e, i, and π. They're all in this one equation, how marvelous is that. It's like the dream team of numbers all in one equation.

This is doubly amazing. First of all it's amazing that there is an equation linking these five numbers. There's no reason why there has to be and it's amazing that it follows so readily from Euler's identity. This equation is so famous that someone wrote a poem about it. W.C. Willig wrote a poem, which I'm now going to recite to you about this equation. Says Willig:

$e^{i\pi} + 1 = 0$
Made the mathematician Euler a hero.
From the real to complex,
With our brains in great flex
He led us with zest but no fearo.

That's pretty great. What else? There's the Euler polyhedral formula from 1752. It says that $V + F = E + 2$ is the way you see it, but I've got to explain what all these letters mean. Maybe I first better say what's a polyhedron? It's a solid body whose faces are planes, plane figures. For instance think of a cube. It's a 3-dimensional body whose faces are square. Now V is the number of vertices of the solid, the number of corners. F is the number of faces on the solid and E is the number of edges forming this. For my cube V, how many vertices are there? There's 1, 2, 3, 4 around the top, 4 around the bottom, so V is 8. How many faces? Think of a die, which has 6 faces, 6 numbers, so F is 6. How many edges on a cube? There are 4 around the top, 4 around the bottom, 4 vertical, 4, 8, 12, and $8 + 6$ is $12 + 2$. The formula works for cubes, but what Euler recognized is it works for this vast range of other solids, other polyhedral like on an icosahedron, like a dodecahedron, like a pyramid.

It's incredibly pervasive in the realm of solids. This formula holds. It's so amazing that this was voted the second most beautiful formula of all time in that poll I mentioned. People think very highly of this. Euler himself was a little more circumspect. Whenever he looked at it he said, "I find it surprising that these general results in solid geometry have not previously been noticed by anyone, so far as I am aware." But, of course, Euler was special in what he could see.

Another great result of his is called the Basel Problem. This was a challenge problem from Jakob Bernoulli in 1689 and the question was can you find the exact sum of this infinite series, 1 + 1/4 + 1/9 + 1/16 + 1/25; it's an infinite series. Notice what the reciprocals are, the perfect squares, the denominators are 2/2, which is the 4, 3^2 is 9, 4^2 is 16. Suppose you take the squares flip them over, add them up, what do you get? Jakob Bernoulli couldn't figure it out. He challenged the world. A generation later Euler comes along and finds the answer to be exactly $\pi^2/6$. This is a bizarre result. This will be the focus of my next lecture. We'll see why it's $\pi^2/6$ and it's phenomenal. This is the result that made Euler famous. He was still a very young man when he did it and in that poll of the most beautiful theorems of all time this is number five. Euler has one, two, and five locked up. We'll see this one again in the next lecture.

What else? There's something called the Euler Path. This comes from a paper called "The Solution" pertaining to a problem of the geometry of position, Geometriam Situs, the Geometry of Position. It was suggested to Euler by the arrangement of the Bridges of Königsberg. Königsberg is a city in Germany. Here is a map from one of his papers about how the city is configured. You see the river flowing through, the river splits, it circles that island labeled A, flows off, and there are bridges connecting the various land masses and you can see those on the picture. According to the story the citizens of Königsberg on Sunday would go strolling around in the park here and across the island. They wanted to take a journey so that crossed each bridge once and only once. This was the challenge. They tried it and it never worked. Either they'd miss a bridge or they'd find themselves crossing a bridge they'd already crossed.

They wanted to know can you do this? Shall we just work harder at it or is this impossible. They asked the Mayor, I don't know why they asked the Mayor, but they asked the Mayor and the Mayor of Königsberg had no idea what's going on. He writes to Euler and Euler answers the question. That in fact with this configuration of bridges it is impossible to cross each path, each bridge once and only once, such a crossing is now called an Euler Path. But, this configuration doesn't have one. He showed other configurations that do and thereby started a subject, which we now know as graph theory.

It's pretty impressive work, but Euler said, "… this solution bears little relationship to mathematics and I do not understand why to expect a mathematician to produce it rather than anyone else for the solution is based on logic alone." He showed it was impossible to take an Euler Path here just using logic. He thought of it as a puzzle. He didn't need math he said because there wasn't any trigonometry in it or anything of that sort. Today this would definitely be recognized as mathematics.

Then there's geometry. He worked in geometry. You think geometry had been around for thousands of years; surely there was nothing left to discover in the realm of geometry, but he did. He found something called the Euler Line of a triangle. Let me tell you about this. Suppose you have any triangle. You can consider the points where the altitudes meet, the point where the altitudes meet. An altitude, remember, is a line from a vertex perpendicular to the opposite side. If you draw the three altitudes, they all meet in a single point called the orthocenter.

You can consider the point where the medians meet. Median goes from each vertex to the middle of the opposite side and they all go through a point called the centroid. You can look at the point where the perpendicular bisectors meet. You take each side bisect, put up a perpendicular, and those three meet at a point. I'm going to show you a picture in a minute. That's called the circumcenter. It's the center of the circumscribed circle. Every triangle has these three points. Let me show you a picture. There's your triangle. Let's first look at the intersection of the altitudes. I draw perpendiculars to each side, there they are, and they all go through a point. There it is; it's called the orthocenter. Let's get rid of those. Let's take a look at the intersection of the medians. Remember the median goes from each vertex to the middle of the opposite side, so there they are and they meet at a point. That's called the centroid. There's that point. Get rid of that. Look at the perpendicular bisectors of the three sides. We bisect each side and put up a perpendicular and they meet at a point called the circumcenter.

Guess what? Can you see it? These points always line up. They always are in a straight line, which is now called the Euler line because he's the first person that saw this in the 18th century. Euclid missed it, Archimedes missed it, Heron missed it—all the great geometers hadn't seen this. It's a fascinating little result and furthermore he showed that the centroid is always half as far from the circumcenter as it is from the orthocenter. In other words those two segments you see in the line are always in the ratio of 1:2 no matter which triangle you look at. It's a great piece of geometry.

Euler also worked in applied mathematics. Here's a diagram from one of his papers where he's studying the theory of machines and he has volumes and volumes and volumes on applied math and there's this little thing that shows up in one of his papers. If you look at this picture of these little circles within circles intersecting circles you probably know this as a Venn diagram except Venn lived in the 19th century and I've taken this picture from something Euler did in the 18th century.

He's already drawing the little circles getting the logical connections among ideas. Probably this should be called an Euler diagram if there were any justice in the world, but truth to tell Venn needs the publicity more than Euler does so we'll still call it a Venn diagram. The Swiss some years ago on their 10 franc note put a picture of their favorite son Euler and here's the bill that had him on it, but I believe it's no longer in circulation, but at least for awhile a mathematician was on the Swiss money even as Newton was on the British pound. It's a nice honor to Euler.

Condorcet, the great French mathematician, said this: "All celebrated mathematicians now alive are his disciples: there is no one who is not guided and sustained by the genius of Euler." I think that is true. I think that remains the case to this day. We're all his disciples. He is the master of us all and I'll leave with a picture of Euler from later in life decked out in his finery and say, "Way to Go, Uncle Leonhard!

Euler's Extraordinary Sum
Lecture 19

> [This] quotation I don't have a source for, but I think it's apt. Somebody said, "*Talent* is doing easily what others find difficult. *Genius* is doing easily what others find impossible." By that definition, Euler—solver of the Basel problem—was indeed a genius.

In this lecture, we look at Euler's solution to the Basel problem, issued in 1689 by Jakob Bernoulli. The challenge was to find the exact value of the infinite series $1 + 1/4 + 1/9 + 1/16 + 1/25...$, the sum of the reciprocals of the squares. Bernoulli himself was able to get an approximation of the sum of this series: a value that is less than 2. In the 1730s, Euler found a more accurate approximation for the series: $1.644934....$ He almost stopped there but later wrote, "... against all expectations I have found an elegant expression for the sum of the series. ..."

To follow Euler's thinking, we need three preliminaries. First, we have to know: For which values of x is the sine of x equal to 0? Delving a little into trigonometry, we find $\sin x = 0$ when $x = \pm 180$ degrees or ± 540 degrees. Converting to radians, we get: The sine of x is 0 if $x = 0$ radians $\pm \pi$ radians $\pm 2\pi$ radians, 3π radians .... The second thing we need is Newton's infinite series for the sine of x: $x - x^3/3! + x^5/5! - x^7/7!$ .... The final piece of the puzzle is from algebra: Suppose $P(x)$ is an infinite-degree polynomial. Suppose further that $P(0) = 1$ and $P(x) = 0$ have infinitely many solutions: $x = a, x = b, x = c$, and so on. Euler was perfectly comfortable factoring $P(x)$ as follows: $(1 - x/a)(1 - x/b)(1 - x/c)$ .... This type of factorization works for second-, third-, and fourth-degree polynomials; Euler extended it to a polynomial of infinite degree.

With this solution, Euler became famous around the world.

To evaluate the series in the Basel problem, we begin by introducing an infinite-degree polynomial: $P(x) = 1 - x^2/3! + x^4/5! + x^6/7!$ .... This is similar to the sine series but not exactly the same. We then factor this polynomial using the preliminaries. We find the factorization works with $P(0) = 1$. What about the solutions to $P(x) = 0$? If we multiply the denominators and the numerators in the polynomial by x, we get Newton's series for the sine of x. By trying to solve $P(x) = 0$, we're now saying that the sine of x/x should be 0. Recall that the sine of $x = 0$ when $x = 0 \pm \pi \pm 2\pi \pm 3\pi \ldots$. Removing 0 from the possible solutions, we're left with $\pm \pi \pm 2\pi$ and so on. We now factor $P(x)$ by the third preliminary, which results in $(1 - x/2\pi)(1 - x/-2\pi)(1 - x/3\pi)$ $(1 - x/-3\pi) \ldots$. Cleaning that up a bit, we see that $P(x)$ is now expressed on the left side of the equation as an infinite sum and on the right as an infinite product. We can link the expressions by multiplying these binomials together two at a time. We now have $P(x)$, after it has been factored and multiplied back together, written as $1 + x^2[-1/\pi^2 - 1/4\pi^2 - 1/9\pi^2 \ldots]$. We equate the coefficients of x^2 in this expression and set them equal, then eliminate the negative signs. Solving, we find that the sum of the infinite series $1 + 1/4 + 1/9 + 1/16 \ldots$ is $\pi^2/6$. With this solution, Euler became famous around the world. ■

Suggested Reading

Dunham, *The Calculus Gallery*.

———, *Euler*.

Edwards, *The Historical Development of the Calculus*.

Euler, *Introduction to Analysis of the Infinite*.

Grattan-Guinness, *From the Calculus to Set Theory*.

Sandifer, *How Euler Did It*.

Questions to Consider

1. As we've seen, Euler proved that $1 + \frac{1}{4} + \frac{1}{9} + \frac{1}{16} + \frac{1}{25} + \frac{1}{36} + \cdots = \frac{\pi^2}{6}$. But he did more. For instance, he found the value of the series of reciprocals of the *odd* squares, as follows: Let $S = 1 + \frac{1}{9} + \frac{1}{25} + \frac{1}{49} + \frac{1}{81} + \cdots$ be the series whose value we seek. By splitting the *original* series into odd and even components, Euler reasoned that: $\frac{\pi^2}{6} = 1 + \frac{1}{4} + \frac{1}{9} + \frac{1}{16} + \frac{1}{25} + \frac{1}{36} + \cdots = \left(1 + \frac{1}{9} + \frac{1}{25} + \frac{1}{49} + \cdots\right) + \left(\frac{1}{4} + \frac{1}{16} + \frac{1}{36} + \frac{1}{64} + \cdots\right)$, so $\frac{\pi^2}{6} = S + \frac{1}{4}\left(1 + \frac{1}{4} + \frac{1}{9} + \frac{1}{16} + \frac{1}{25} + \cdots\right) = S + \frac{1}{4}\left(\frac{\pi^2}{6}\right)$ (why?). From this, deduce, as Euler did, that $1 + \frac{1}{9} + \frac{1}{25} + \frac{1}{49} + \frac{1}{81} + \cdots = \frac{\pi^2}{8}$.

2. Euler also examined the alternating series of reciprocals of squares: $1-\frac{1}{4}+\frac{1}{9}-\frac{1}{16}+\frac{1}{25}-\frac{1}{36}+\frac{1}{49}-\frac{1}{64}+\cdots$ and showed that this summed to $\frac{p^2}{12}$. Use question 1 to show how he got this answer.

Euler's Extraordinary Sum
Lecture 19—Transcript

In our previous lecture, we met Leonhard Euler, the master of us all. We surveyed some of his work, a very small fragment I might say. In this lecture and the next, I want to show you his mathematics. I want to examine a great theorem from Euler in each of these lectures. These are spectacular, but they are heavy into the mathematics. I reiterate my promise that I will try to take the material and break it down into understandable steps so that you too can follow along as he does this extraordinary work. These theorems, this one and the next, I think are as spectacular as anything in the course. I think they're a real treat. Let's begin.

The theorem for this lecture is his solution of the Basel problem, which I mentioned previously. The problem was a challenge from Jakob Bernoulli. It was issued in 1689. He published this as a challenge to the mathematical world. It was to find the exact value of the infinite series $1 + 1/4 + 1/9 + 1/16 + 1/25$. You see what this is. We're taking the whole numbers, squaring them, taking their reciprocals, adding them up forever—this is another one of these infinite series problems that was of such interest to people back then—and figure out what is the sum exactly. Jakob couldn't do it.

We saw in an earlier lecture, that if you take the triangular numbers that you can make into little triangles—$1 + 1/3 + 1/6 + 1/10$—Leibniz figured out that sum was 2. That wasn't too hard. Here I'm taking the square numbers, a slightly different geometrical configuration. This one is way harder. To Jakob's credit, he did get an estimate of the answer. He could get a bound on the answer. Let me show you what he did here to get at least an inkling as to what this series would sum to. There's the series—$1 + 1/4 + 1/9 + 1/16$—the sum of the reciprocals of the squares. I'm going to follow Jakob along in bounding this or approximating this.

Here's what you do. Write this out as $1 + 1$ over 2×2. Instead of 1/4, split it up—2×2. The 1/9 becomes $1/3 \times 3$, the 1/16, $1/4 \times 4$ and so on. That's the same series. Now what I'm going to do is compare that to a series where I slightly modify the denominators. Let me put that series up. It starts off at the 1, but then the next term is $1/1 \times 2$ instead of its companion right above

$1/2 \times 2$. When I move to the next one instead of $1/3 \times 3$ I change it to $1/2 \times 3$. Then comes $1/3 \times 4$, $1/4 \times 5$ and the question is how does that new series on the lower line compare to the original series above?

You just think about it a minute. Look up above, I start with a 1 in each so those are equal, but if I move to the next one I have $1/4$ up above, $1/2 \times 2$ and $1/2$ down below. You think about it and you see a $1/4$ is less than $1/2$ so the top is less than the bottom. Move to the next one $1/9$ on top, a $1/6$ down below, a $1/9$ is less than $1/6$. A $1/16$ is less than $1/12$. All the ones above are less than all the ones right below them. The inequality sign I want to put here is a less than. That is the top series; the one I'm after is less than the one below.

The one below if we multiply this out is $1 + 1/2 + 1/6 + 1/12 + 1/20$ and Bernoulli could handle this. He would say that we'll leave the 1 in the front, but then when you get to that $1/2$ write it as $1 - 1/2$. When you get to the $1/6$ write it as $1/2 - 1/3$. The $1/12$ is a $1/3 - 1/4$, the $1/20$ is $1/4 - 1/5$. We saw Leibniz employ this very same trick in summing a series the challenge problem from Huygens. It's a trick that's kind of familiar and now what happens is you get cancellations up and down the line.

The $-1/2$ takes the $+1/2$ beside it. The $-1/3$ takes the $+1/3$, the $-1/4$ takes the $+1/4$, and so on. Everything cancels in the tail of the series. All that's left is the 1 in the parenthesis and there was that 1 at the front end. If you look at what remains we get 2. The series in question is less than 2 and Bernoulli knew that. He stated that whatever the answer to this is, whatever $1 + 1/4 + 1/9$ and so one comes out to be it's really less than 2. But, that wasn't good enough. He wanted to know exactly what it is and he couldn't do it and he wrote this in his challenge.

He said, "If anyone finds and communicates to us that which thus far has eluded our efforts, great will be our gratitude." We'll really be thankful if you can give us the answer. We really want to know. This was dangerous for him to write this of course because his brother Johann might've done it and then he could've come running in you know waving the solution in Jakob's face, but I'm sure he would've enjoyed that, but Johann couldn't do it either. This was hard. The problem remains out there for a generation known as the Basel Problem because Jakob had issued it from Basel, Switzerland. Whoever did this is going to be famous, but nobody could.

In the 1730s, Euler found an accurate numerical approximation for the series. Using some rather clever techniques he found that the series in question, the 1 + 1/4 + 1/9 comes out somewhere near 1.644934 and that was pretty accurate. If that had come out to be I don't know something like 1.6666667 you'd have a clue as to what the answer should be. That looks like a decimal for 5/3, but 1.644934 doesn't look like anything to me so it's not clear that this helped and Euler said, "It seemed most unlikely to be able to find anything new about this." It seemed like that was as far as we were going to get. Ahh, but finally he said, "Against all expectation I have found an elegant expression for the sum of the series." He almost gave up on it too, but then he saw something and that's what I want to show you, what he saw and how he figured out the value of this series exactly.

To do this you need three preliminaries, three lemmas I guess we could call these, three things in the background. Let's take a look at these one at a time and then you blend these together with just the right touch of Euler and genius and you can figure out what the series equals. The first one is this. For which values of x is the sin of x equal to 0? That's going to become important in what follows. Now we've got to kind of delve into the realm of trigonometry and remember what the sine is and see if we can figure out the x's that make the sin of x 0.

Here's a diagram. Remember we're going to be working in the unit circle so the radius is 1 from O to A is 1. Let me let x be the size of the angle opening there. Angle AOB has measure x. If you drop BC perpendicular to OA then you've got to form this triangle here, triangle OBC, and in that triangle you can see the sin showing up. The sin, remember, is that trigonometric ratio opposite side over hypotenuse. In the diagram the sin of x, the sin of the angle x is the opposite side over the hypotenuse, but the opposite side is BC and the hypotenuse is OB, so it's the ratio of BC to OB, but OB the radius of that circle is 1. This is a unit circle and so the sin is BC over OB, that's really just BC over 1, which is really just BC. In the picture the sin of angle x is BC, which if you look at it is that vertical distance from point C up to the circle wherever you end up when you move through the angle. There's the picture again. I say the sin of x we said is BC, the vertical distance, and so if I ask you when is the sin of x equaling 0, what I'm really asking is when is BC equal to 0, when is that vertical distance 0.

If you'll think about this for a minute, suppose x is 0 degrees, suppose x had measure 0. You don't start going around yet. You're just running around the line OA; the whole angle just goes from A to O and back to A again, 0 opening. Then BC is certainly 0, it's the height up to the circle, but there is no height if your angle is 0. As you spin your angle around, the BC will rise as it reaches the crest and then fall back again and it will be 0 again, BC will be 0 again when you're 180 degrees away from A. In other words, you're at the other end of that diameter, at the left end. Then you're going to hit another 0 value for the sin.

Then you sin if you continue around in a counterclockwise direction, you'll get these negative values of the sin, the BC is going down until they come back up and you're at the right end of the diameter and you get a 0 again. If you think about it the sin will be 0. Whenever your angle leaves you at the right end of that diameter or at the left end and then you just have to figure out when that is, which angles do that. If you had 0 degrees you're over at A, the sin is 0. If you're at 180, you're at the other end of the diameter the sin is 0, but you can actually go backwards too. You might remember we can measure angles clockwise or counterclockwise. If I go the other way −180 will bring me to the same point. I'm going to say that the sin of x is 0 when x is ±180 degrees.

Or, you could do a whole revolution ± 360, either clockwise or counterclockwise, you'll get yourself back to A again, or ± 540, now you'll be back over on the other side. These are the angles, the values of x that can give you a sin of $x = 0$, x out. These are in degrees and remember when we're doing higher mathematics we don't want degree readings we want radian measure. I've got to convert all of these to radians. That's easy enough to do. Remember how the radians work. You're measuring the length of the arc AB, that's the size of your angle.

If you do a whole lap around you go 2π, $2\pi^r$, but the radius of the circle is 1 so a 360-degree rotation is equivalent to 2π radiation rotation, 180 degrees is 1π and so here goes my solution. This is what I really want here. The sin of x is 0 if x is 0 radians $\pm \pi$ radians $\pm 2\pi$ radians, 3π radians and so on. You can see this if you graph the sin wave. If you just graph the function $y = \sin x$, there it is. You certainly have seen this wavy thing and look where it's crossing the x-axis. It's 0 at π, at $-\pi$, at 2π and -2π, 3π and -3π and so my first preliminary is if the sin of x is $0 \pm \pi \pm 2\pi \pm 3\pi$ et cetera. That's the first thing we're going to need to store away. We'll be back to that in a minute.

The second thing we need to proceed with Euler's derivation is Newton's infinite series for the sin of x. We talked about this in a previous lecture. Newton discovered this amazing fact if the sin of x where x is in radians is $x - x^3/3! + x^5/5! - x^7/7!$ and so on. I mentioned that this is a very beautiful formula, nicely structured, very regular in its terms. Newton had discovered it, Euler certainly knew it, and he's ready to use it. That's the second piece of the puzzle.

The third piece is a fact from algebra and again this'll take a little bit of time just to see what's up here. But, let me start simple here, suppose we have a second-degree polynomial, the quadratic, and suppose you know a few things about it. You know that $P(0)$ is 1. If I call my polynomial $P(x)$ and if I put 0 in I get 1 out. Suppose you also know that $P(x) = 0$ has two solutions $x = a$ and $x = b$. Okay, so that is the given. Then, you can factor $P(x)$ is follows. You can write $P(x)$ is the product of two things $(1 - x/a)(1 - x/b)$.

Why is that? Let's see if we can understand this. First of all look at that equation $P(x)$ is $(1 - x/a)(1 - x/b)$. If you were to multiply this back together you'd surely get a quadratic, a second-degree, there'd be an x^2 in it, nothing higher. For sure that's a quadratic polynomial there. What if you put 0 in? What is $P(0)$ and if I put 0 into this factored formula? According to the formula $P(x)$ is $(1 - x/a)(1 - x/b)$. So $P(0)$ is $(1 - 0/a)(1 - 0/b)$, just letting x be 0, but $(1 - 0/a)$ is 1, $(1 - 0/b)$ is 1 and so $P(0)$ is 1×1, which is what it was supposed to be. $P(0)$ is supposed to be 1. That checks out. What if I put $x = a$ into this? What if I asked for what's $P(a)$? Now $P(a)$ will be $(1 - a/a)$ $(1 - a/b)$. But, if you look at that a/a is 1 so the $P(a)$ is $1 - 1$ times the other term $(1 - a/b)$. Well $1 - 1$ is 0 and once you've got a 0 you multiply you get 0. Sure enough with that factored form $P(a)$ is 0 and you can see that $P(b)$ is 0 for the same reason. I guess the point is that the factorization I showed you

$P(x) = (1 - x/a)(1 - x/b)$ meets the criteria. It's a quadratic. $P(0)$ is 1, $P(a)$ is 0, $P(b)$ is 0, that's how you factor it. That wasn't too hard.

Let's crank it up. Suppose you have a third-degree polynomial, third-degree. You know $P(0)$ is 1 again and you know that $P(x) = 0$ has three solutions now, $x = a$, b, and c. I want to factor $P(x)$ now. It's exactly analogous to the previous case. The factorization now has three terms $(1 - x/a)(1 - x/b)$ and then you stick on $(1 - x/c)$ and you can check that's a third-degree polynomial—$P(0)$ is 1, $P(a)$ is 0, $P(b)$ is 0, $P(c)$ is 0, sure. That's not bad.

Let me crank it way up. Suppose $P(x)$ is an infinite-degree polynomial. Euler would always deal with these things. It just goes on forever. Suppose $P(0)$ is 1 and suppose $P(x) = 0$ now has infinitely many solutions, $x = a$, $x = b$, $x = c$, etc. Euler was perfectly comfortable. Generalizing the previous results and saying here's how you factor $P(x)$. You can write it as $(1 - x/a)(1 - x/b)$ $(1 - x/c)$ and the product goes on forever, infinitely many factors, there you go. The fact that this works for second-degree and third-degree, and fourth-degree Euler says let's just extend it to a polynomial if you will of infinite-degree, just one that never ends.

Those are the three preliminaries; now he's ready to put them together and solve the Basel Problem. In order to evaluate this series, which remember was $1 + 1/4 + 1/9 + 1/16 + 1/25$, he's going to begin by introducing an infinite-degree polynomial, well-chosen to be this. This is what he wants to look at. $P(x) = 1 - x^2/3! + x^4/5! + x^6/7!$ and so on. We haven't seen this one before although it sort of calls to mind maybe the sin series, but it isn't, the powers aren't right. In any event it's going to be an infinite-degree polynomial $P(x)$.

What Euler wants to do is factor it using these preliminaries. He wants to see if I can break this up into a product of pieces and we can if we use the result I just showed you if it applies to this $P(x)$. The first thing he's got to check is $P(0)$ 1. If we're going to use that result I just saw we always demanded that $P(0)$ be 1. Look at it. Put 0 in there. $P(0)$ is $1 - 0^2/3! + 0^4/5!$, you put in 0 for x, you're going to wipe out everything except that lead 1. Sure enough $P(0)$ is 1. That was one of the things required to do that factorization so we check that off. He's got that under control.

Next up, what are the solutions to $P(x) = 0$? Remember these are going to be like the a's and the b's and the c's that we're going to put into our factorization so we've got to solve $P(x) = 0$. This looks really tough. It looks like we're going to go off the rails here because $P(x)$ is this monstrosity here. $1 - x^2 3! + x^4/5!$, etc. It's got infinitely many terms. I set this equal to 0 and I've got to figure out what x is. I've got to solve this. Yikes, what do you do?

Here's what you do. Euler says we can do this. There's $0 = P(x)$, there it is $1 - x^2 3! + x^4/5!$. I'm trying to solve this, set this equal to 0. Draw a horizontal line. In the top and bottom I'm going to multiply by x. I hit the top with an x, I hit the bottom with an x. I can do that. You don't change your fraction by doing the same thing to the top and bottom. That's still equal except for one little snag here, x can't be 0 here because then I would have a 0 on the bottom and you're never allowed to have that. This is a legitimate operation provided x isn't 0.

Now on the numerator multiply through, that x multiplies each term, and so you end up with 0 on the left and on the right you've got that x in the denominator, but up on top when I hit an x across each term. I get $x - x^3/3! + x^5/5! - x^7/7!$. We just increased each power up there by 1. But wait a minute that numerator I hope looks familiar. That's Newton's series. That's the sin of x. So I'm trying to solve $P(x) = 0$ and by this little trick of multiplying top and bottom by x I'm down to saying that the sin x/x should be 0.

That I can solve because if you just cross multiply here the sin of x would be 0 and now you've got to remember when is the sin of x 0. That was our first preliminary; we saw that. The sin of x is 0. We said precisely when x is $0 \pm \pi \pm 2\pi \pm 3\pi$ and so on except there's one little catch here. Remember we know that $P(0)$ is 1 so $x = 0$ is not a solution to $P(x) = 0$ because $P(0)$ isn't 0. $P(0)$ is 1, we've got to toss that out, because 0 is not a solution to $P(x) = 0$. Where did it come from? It sort of sneaked in there. Remember when I multiplied top and bottom by x, I said x couldn't be 0, that would be illegal, but then I sort of forgot about it when I said sin of x is 0, I threw 0 into the pot, 0 has to be removed.

Your only solutions are $\pm \pi \pm 2\pi$ and so on. Those are like the a's, the n's, and the c's when I try to factor P(x). We're getting near the end here and so take a deep breath. P(x) is an infinite-degree polynomial. P(0) is 1, the solutions to P(x) = 0 we know, $\pm \pi$, 2π, 3π.

Here comes the factorization by that third preliminary. First I'm going to write down what P(x) was just to remember. It's $1 - x^2 3! + x^4/5!$. That was P(x). Now I'm going to factor it. The preliminary said you take $1 - x/a$ where a is the first solution to your equation. That would be π. We said that the solutions were plus or minus π. Let's put π in there, $1 - x/a$ is like $1 - x/\pi$. Then the next one will do the minus π, so $1 - x/b$ is like $1 - x/-\pi$. Then we got and get $(1 - x/2\pi)(1 - x/-2\pi)(1 - x/3\pi)(1 - x/-3\pi)$ and you've got this infinite product of these factors.

Euler wants to clean this up a little bit. It starts off with $(1 - x/\pi)$; I'll change the second one to $(1 + x/\pi)$, just get rid of the double negative. The next one is $(1 - x/2\pi)$, that's fine. The fourth term in the product $(1 - x/-2\pi)$ we turn that into $(1 + x/2\pi)$ and then there's $(1 - x/3\pi)$ and $(1 + x/3\pi)$. On it goes an infinite product. P(x) is now expressed on the left side of the equals sign as an infinite sum and on the right side of the equals sign is an infinite product. One of Euler's tricks, how he would do his amazing mathematics was he would try to express the same thing in two very different ways. Look here we've got P(x) expressed as a series and as a product, same thing, P(x) maybe by linking these together we can get somewhere and that's exactly what he does.

Next up, we're going to multiply these binomials together two at a time. Look at the first two, $(1 - x/\pi)(1 + x/\pi)$. Those could be multiplied together and come out to be $(1 - x^2/\pi^2)$ sort of like $(a - b)(a + b)$. The next two, $(1 - x/2\pi)(1 + x/2\pi)$, that multiplies to be $(1 - x^2/4\pi^2)$. The next two will multiply to be $(1 - x^2/9\pi^2)$. Now look at the denominators here. I'm starting to see something. I see a 4 and a 9, the next one would be $(1 - x^2/16\pi^2)$, somewhere here I'm getting an inkling that he's on to something and he's getting these squares showing up in the bottom. Remember that was what the whole problem was to sum the $1 + 1/4 + 1/9$. We're not there yet, but you begin to see he's thinking.

There we have it, P(x) written as the infinite series at first and then this product of these terms all of which have x^2 in them. What Euler says is multiply that out, all those things in the square brackets, let's multiply that out. If you do this you always have to take one piece of each thing in the square bracket and multiply along and you can see what the constant term will be, 1 comes out of the first bracket times 1 from the second square bracket times 1 from the third. You multiply all those 1's you get 1. It's going to start with a 1.

How many x's will emerge from this? If you multiply all those terms in the square brackets you can never get an x because you always have x^2 or you could have x^4 or x^6 as they multiply. You can never get an x^1 so there's not going to be any x's. How many x^2 are going to come out if you multiply all these square bracket terms together? There's going to be some of them, and what Euler says is think of where the x^2 will come from.

Suppose you take that first square bracket where you have a $-1/\pi^2(x^2)$, suppose that $-1/\pi^2$ multiplies all the other 1's, you'll get a $-1/\pi^2x^2$. L:ook at the second square bracket, you have a $-1/4\pi^2(x^2)$. That $-1/4\pi^2$ can multiply all the other 1's and you'll get a $-1/4\pi^2x^2$. From the third term you'll get a $-1/9\pi^2x^2$ and so on down the road. If you factored out the x^2 what you'd be left with is the coefficient of the x^2 is a $-1/\pi^2 - 1/4\pi^2 - 1/9\pi^2 - 1/16\pi^2$ and so on. That would be your x^2 coefficient. After that I don't care what happens. Euler just says you know whatever happens thereafter we don't need to worry about it. I'll just say stuff, more stuff comes out as you multiply and you get x^4 and x^6. For this problem that's irrelevant.

Where are we? We know that P(x), which is $1 - - x^2/3! + x^4/5!$ And so on has now been written after I've factored it and then multiplied it back together as $1 + x^2[-1/\pi^2 - 1/4\pi^2 - 1/9\pi^2$ and so on, plus whatever's out there on the end. You take a deep breath and equate the coefficients of x^2 in this expression. P(x) has been expressed as a series in two different ways. Notice they both start with 1, both to the left and the right of P(x). Then you come to the x^2. How many x^2 in the first expression $-1/3!$. How many x^2 on the right-hand expression? That's what's in the square brackets $-1/\pi^2 - 1/4\pi^2 - 1/9\pi^2$. Those are the coefficients of x^2 for p(x). Those must be the same and so you set them equal.

Now let's get rid of all the negative signs. We'll just turn all the negatives to positive, everything is negative there, turn it into a positive 1/3!, 3! Is 3 × 2 × 1, that's 1/6. On the right side pull out the $1/\pi^2$ and you're left with 1 + 1/4 + 1/9 + 1/16 inside the square brackets is the very thing Euler's after. He has to sum that series, but he's got it now because 1/6 is equal to $1/\pi^2$ times the series, cross multiply by π^2 and you get that the sum of the infinite series 1 + 1/4 + 1/9 + 1/16 is sure enough $\pi^2/6$ and with that argument Euler solved the Basel problem. That's pretty good!

When he published this, when he had solved this great unsolved challenge, he became famous around the world. Note that the answer $\pi^2/6$ is less than 2 as Bernoulli had said. Bernoulli was right when he approximated it or he bounded it and it's actually pretty close to 1.644934 as Euler had also determined, but this is better than either of those because this gets it exactly. Euler did it.

I want to end this with two quotations that I think are germane. One is from Ivor Grattan-Guinness, a 20th-century math historian who said, "Euler was the high priest of sum-worship," a play on sun-worship of course, "for he was cleverer than anyone else at inventing unorthodox methods of summation." Euler could do this sort of thing better than anyone. The other quotation and I don't have a source for, but I think it's apt, somebody said, "*Talent* is doing easily what others find difficult. *Genius* is doing easily what others find impossible." By that definition, Euler solver of the Basel Problem was indeed a genius.

Euler and the Partitioning of Numbers
Lecture 20

In 1750, ... [Euler] writes a paper called *De Numeris Amicabilibus* on amicable numbers, and in this paper, he finds 58 more pairs. The supply had gone from 3 to 61. He multiplied the number of known amicable pairs by 20 in one paper.

Before we look at this lecture's great theorem, we briefly survey number theory. Among Euler's achievements in this area was his discovery of 58 pairs of amicable numbers, that is, whole-number pairs in which each is the sum of the proper whole-number divisors of the other (e.g., 220 and 284). Until Euler explored amicable numbers in 1750, only three such pairs were known. Euler also answered the challenge of finding four different whole numbers, the sum of any two of which is a perfect square: 18,530; 38,114; 45,986; and 65,570.

> **Euler also answered the challenge of finding four different whole numbers, the sum of any two of which is a perfect square: 18,530; 38,114; 45,986; and 65,570.**

Our great theorem for this lecture relates to the partitioning of numbers. To start, let $D(n)$ be the number of ways of writing the whole number n as the sum of distinct whole numbers. For example, $D(5)$ would be 3: 5; 4, 1; and 3, 2. Now let $O(n)$ be the number of ways of writing n as the sum of odd numbers that are not necessarily distinct; $O(5)$ would be 3 again: 5; 3, 1, 1; and 1, 1, 1, 1, 1. The values $D(8)$ and $O(8)$ are both 6. In fact, Euler found that for all whole numbers, $D(n)$ and $O(n)$ are always the same.

Euler broke the proof of this theorem into three parts. First, he introduced $P(x)$, which is $(1 + x)(1 + x^2)(1 + x^3)(1 + x^4)(1 + x^5)...$, an infinite product of binomials. $P(x)$ is equal to 1 plus the sum as n goes from 1 to infinity of a certain number of x^n's. How many x^n's? Exactly $D(n)$, that is, exactly the number of ways of decomposing n into distinct pieces.

The second part of the proof involves an infinite geometric series: $1 + a + a^2 + a^3$.... The sum of this series is $1/1 - a$. In his proof, Euler introduced $Q(x)$, which is $(1/1 - x)(1/1 - x^3)(1/1 - x^5)(1/1 - x^7)$. Note that the powers here are odd. To eliminate the denominators, we replace the $1/1 - x$ by a series using the formula for the infinite geometric series; thus, $1/1 - x$ becomes $1 + x + x^2 + x^3$..., an infinite series. The next expression is $1/1 - x^3$. We replace this fraction with the infinite series $1 + x^3 + x^6 + x^9$.... At this point, we have infinitely many infinite series.

We now rewrite the first expression, $1 + x + x^2 + x^3$..., as $1 + x^1 + x^{1+1} + x^{1+1+1}$.... The second expression, $1 + x^3 + x^6 + x^9$..., becomes $1 + x^3 + x^{3+3} + x^{3+3+3}$ .... $Q(x)$ is equal to the product of all these expressions. We then multiply out these infinitely many infinite series. Again, we find that $Q(x)$ is 1 plus the sum as n goes from 1 to infinity of a certain number of x^n's. How many x^n's? Exactly $O(n)$, that is, exactly the number of ways of decomposing n into odd summands.

Notice that $P(x)$ is equal to 1 plus the sum of $D(n)x^n$, and $Q(x)$ is equal to 1 plus the sum of $O(n)x^n$. Recall that we're trying to prove that $D(n)$ is always equal to $O(n)$. That would be true if $P(x)$ and $Q(x)$ were the same, but they don't appear to be. Euler showed, however, that they are the same by changing the original $P(x)$, $(1 + x)(1 + x^2)(1 + x^3)$..., to a fraction. Then, canceling terms, $P(x)$ becomes $1/(1 - x)(1 - x^3)(1 - x^5)$..., the original $Q(x)$; thus, $D(n)$ must equal $O(n)$ for all n. ∎

Suggested Reading

Dunham, *Euler.*

Euler, *Introduction to Analysis of the Infinite.*

Ore, *Number Theory and Its History.*

Sandifer, *How Euler Did It.*

Weil, *Number Theory.*

Questions to Consider

1. Find $D(10)$ and $O(10)$, then find $D(13)$ and $O(13)$, where, as in the lecture, $D(n)$ is the number of decompositions of n into the sum of *distinct* whole numbers and $O(n)$ is the number of decompositions of n into the sum of *odd* whole numbers. Needless to say, in both cases, your values should be the same (as Euler proved they must be).

2. As Euler was nearing the end of his career, he shared the mathematical spotlight with Joseph-Louis Lagrange (1736–1813), who gave us the so-called Lagrange four square theorem. This shows that any whole number can be written as the sum of four or fewer perfect squares. The theorem, proved by Lagrange with an assist from Euler, is one of the most intriguing results in all of number theory. Check it for $n = 13$, $n = 28$, and $n = 115$. That is, write each of these numbers as the sum of four or fewer perfect squares.

Euler and the Partitioning of Numbers
Lecture 20—Transcript

In our previous lecture, we saw Euler solve the Basel problem, a great theorem if ever there was one. There he was working on infinite series, which is a subject that had been around for a long time. He was pushing the frontiers. In this lecture, I want to show you another of his great theorems on a subject that was brand new. He was essentially making up a field of mathematics as he went along. It's a theorem about the partitioning of numbers.

However, I have a little extra time in this lecture so I thought I'd start by just surveying one branch of mathematics we haven't mentioned yet and that is number theory. Euler was one of the great number theorists of all time, certainly in that same chain that brought us Euclid and Fermat. You've got to include Euler. He wrote volumes of wonderful mathematics on number theory. Let me just tell you about one of the contributions he made. A definition had been kicking around for centuries. It is as follows. It says whole numbers M and N are amicable, which means friendly, if each is the sum of the proper whole number divisors of the other.

This needs an example, so here's an example. Look at the number M = 220 and N = 284. What I'm looking for are the proper divisors of 220. By that, I mean whole numbers that divide evenly into 220, but are not 220. That's what the proper means, smaller than. If you look at the divisors of 220, you'll get 1, 2, 4, 5, 10, 11, 20, 22, 44, 55, and 110. You also get 220, which divides into 220, but that's not proper. If you add those up you get 284. Look at the proper divisors of 284: 1, 2, 4, 71, and 142. Those are the whole numbers that go into 284 and are less than 284. Add those up and you get 220.

Each of these numbers is the sum of the proper divisors of the other. It's a strange reciprocity. It has no particular use, but it is intriguing. Mathematicians, number theorists, love this. I have a friend who, when he married the love of his life, gave her a keychain with 284. He has a keychain with 220 on it to indicate their undying friendship. It's very sweet.

Let me give you the history of amicable numbers. It's very short. The Greeks knew that pair I just showed you, 220 and 284, the Greeks had somehow discovered and they wanted to find more and they couldn't. These are hard to come by so that was it. They just knew that one pair.

In the 9th century the Islamic mathematician Thabit ibn Qurra's, whom I mentioned in an earlier lecture, found a rule that generated two more pairs. Now there were three known pairs. However, apparently Thabit's rule didn't make it back to Europe after the Renaissance where the European mathematicians only knew of that Greek pair, only that first pair. In the year 1636 Pierre de Fermat, our friend the number theorist, found another pair 17,286 and 18,416. Believe it or not if you add up all the proper divisors of 17,286 you get 18,416 and vice versa, and you can check this if you wish, but that's a pair.

Guess what—that's one of Thabit's numbers. Fermat was rediscovering something that Thabit had found centuries before. With that discovery Fermat's looking pretty good so his rival Descartes had to find a pair of his own to sort of you know show his mathematical prowess. Descartes goes to work on it and in 1638 he finds that pair, 9,363,584 and 9,437,056. I figure if you make it to your 50th wedding anniversary, you give your spouse that number on the keychain and it's really amazing. Guess what, that's the other number that Thabit had found. Fermat and Descartes were really just retracing the ground. These are the three easiest pairs of amicable numbers, believe it or not. These are the ones that come up first and that's what the world knew in 1638, three pairs. A century later three pairs, no one had found anymore.

In 1750, Euler thinks about this. He writes a paper called *De Numeris Amicabilibus* on amicable numbers and in this paper he finds 58 more pairs. The supply had gone from three to 61. He multiplied the number of known amicable pairs by 20 in one paper. This is what Euler would do. He would blow these problems out of the water. It's really quite astonishing, plus I love the word amicabilibus, it's fun to say and it's part of the title of his paper. That's an Euler result in number theory.

Let me mention one other little curiosity I would call this. We're getting to our great theorem in a minute, but this is just prologue. Someone challenged Euler to find four whole numbers, the sum of any two of which is a perfect square. Four whole numbers, if you add up any two you get a perfect square. Wait, we have to qualify this. I could say here they are, 8, 8, 8, and 8 and if I add up any two of them I get 16, which is a perfect square. That's too easy so we've got to say four different whole numbers. There we go. That makes it a lot harder. Four different whole numbers, the sum of any two of which is a perfect square, imagine you're trying to create these yourself. So here's what you might try. You start with 1. Now I've got to find something to add to 1 to make a perfect square, so maybe 3, 1 + 3 is 4, a perfect square, great. Now I've got to find something to add to 3 to make a perfect square so you try 6, 6 and 3 is 9, perfect square, but you can see what went wrong. You just messed up the 1, 6 and 1 is 7, that's not a perfect square. Now you could go back and change the 1 to 10, so the 10 and the 6 is 16, which is a perfect square, but now the 10 and 3 doesn't work.

It's like you're juggling balls and while you're worrying about these two you drop the third one. That's just three numbers. Euler had to find four numbers so the sum of any two is a perfect square. He thinks about it and he said try these four. How about 18530, 38114, 45986, 65570 and you can check it, they work, the sum of any two of those is a perfect square, but when you see these you sort of think he must've been a space alien or something. How does anybody come up with numbers like that? Then you read it and he tells you how he did it and there's of course a method and it's very logical and by the time you're done you say oh I could've thought of that. That wasn't so hard, but you would never have thought of that. This was how Euler worked. That was just a little background.

Here comes my great theorem for this lecture. It's on the partitioning of numbers and it requires some terminology notation and background so again we'll try to go through this very deliberately to see this result. Let me first of all let $D(n)$ be the number of ways of writing the whole number n as the sum of distinct whole numbers, different whole numbers, so the D stands for distinct.

We need an example, so let's just try one here. Suppose n is 5. So what I'm trying to do is break 5 up into the sum of different size pieces. In $D(5)$ will be how many ways to do this. For 5 it isn't too hard. Let's see, we want to break 5 up into distinct pieces. One of them I'll allow is 5 itself, so it's the entire 5, not broken at all. That's going to be one. How about 4 and 1, that makes 5 and they're different, 4 and 1, so that's the second way to do it. How about 3 and 2, that makes 5 and they're different, yeah.

Are there some more? You can say 2 and 3, but that's really the same as 3 and 2, we're not going to count that as a different one. You could say how about 6 and –1, they're different and they add up to 5, which is true, but –1 isn't allowed. These have to be whole numbers, positive. That's no good. You could say, what 3-1/2 and 1-1/2, which add up to 5, they're different, but they're not whole numbers either, so that's no good. If you think about it, these are the only ones 5, 4 and 1, 3 and 2. That's the only way to break 5 up into different size pieces so D(5) remember there's how many ways to do this, 3. There they are, those 3.

Let me let O(n), new symbol, be the number of ways of writing n as the sum of, not necessarily distinct, odd numbers. Now we want to take n and break it up into odd sum ends, but this time I'm going to allow them to be repeated, but you see what I'm restricting is they've got to be odd. Let's look at an example. Let's do 5 again. So I want to break 5 up into odd pieces. One of them could be 5 standing along, how about 3 and 1 and 1—three, four, five—and they're all odd. Notice I repeated the 1 here, but for this game, for the O(n) I'm allowed to repeat as long as they're odd. That's the second way to do it. How about 1 and 1 and 1 and 1 and 1, five 1's they're all odd, that's 5. Then if you think about it you're not going to find anymore. That's it. We would say O(5) has how many ways to split 5 up into odd pieces O(5) is 3 and there they are, those three decompositions.

Before we move on let me do another example. Let's try 8. Let's look at the number $n = 8$ and do this again for 8. If I'm looking for distinct sum ends, now you've got to break 8 up into different size pieces. Here we go, 8 stand alone, great. How about 7 and 1, that makes 8 and they're different, sure, 6 and 2, yep that makes 8, 5 and 3, good that makes 8, 4 and 4, well that makes 8, but they're not different. That's out. We can't count that one. But, there are a few more out there. How about 5 and 2 and 1, that makes 8 and there's three sum ends now, but they're all different so that's good. How about 4 and 3 and 1? That makes 8 and they're all different. If you try to find anymore you won't. That's the whole collection and how many ways to do that? One, two,

265

three, four, five, six, six ways to break 8 up into the sum of distinct sum ends so D(8) is 6. Remember D(8) counts how many ways to do the process, 6.

Let's do O(8), let's break 8 into odd sum and see what happens, so 8 has to get split up into odd pieces, 7 and 1, that's 8, 5 and 3, that's 8 and they're all odd. How about 5 and 1 and 1 and 1? Sure, that's 8. Remember we're allowed to repeat this time as long as they're odd. How about 3 and 3 and 1 and 1? That's 8, how about 3 and five 1's. That's 8 and how about 8 one's, sure. There are the decompositions of 8 into odd sum ends and now you count them up 1, 2, 3, 4, 5, 6. Hmm, O(8) is 6.

Let's summarize. Where are we here? D(5) was equal to O(5), they were both 3. D(8) was equal to O(8), they were both 6, you know, now if you tried this for D(10) and O(10) they'll come out the same. D(12) and O(12), they'll come out the same. Is there something going on here, some theorem? In 1740, Philippe Naudé called Euler's attention to this whole matter of partitioning whole numbers. Naudé was a French mathematician, he sent Euler a letter and said we need to think about this. This is an interesting problem.

Within days, Euler sent back a proof of the following great theorem that I'm going to show you and in the process apologizing for the delay caused by his "… bad eyesight, he wrote, which I have been suffering for some weeks." His eyes were giving him trouble and he apologized that it took him a few days to solve this great theorem. For most people this is a career maker, Euler did it with painful eyes, you know, in a very short period.

Here's the theorem. That this phenomenon we observed with 5 and 8 is true always. For all whole numbers D(n), the number of decompositions of n into distinct pieces and O(n), the number of decompositions of n into odd pieces always are the same. Euler is going to prove this for all whole numbers at one. One proof will get it for all n's, one shot, you know one proof fits all. How in the world do you do this? He breaks it into three pieces. All we've got to do is follow through these three pieces and we'll see the result emerge as if by magic. It's really spectacular. First piece, he says I'm going to introduce this thing. P(x) we'll call it, is $(1 + x)(1 + x^2)(1 + x^3)(1 + x^4)(1 + x^5)$ forever, an infinite product of these binomials.

Why is he doing this? Just hang on and you'll see. But, that's what he's going to introduce P(x). He says let's multiply it out. Multiply out this infinite product. Let's see, what's going to happen? First of all, what's the constant term going to be here? The constant will arise if you multiply 1 from the first binomial times the 1 in the second, times the 1 in the third, $1 \times 1 \times 1$ starts off with a 1.

How many x's are you going to get when you multiply this out? There's that x in the first term and it can hit all the other 1's so you'll get an x, but if you look at the rest of this expression for $P(x)$ you'll see there's no other source of an x. You can never get another x^1 except that 1. This is going to start off $1 + x$. How about x^2? The x^2 in the second binomial could multiply all the other ones and give you an x^2, but there's no other way to get an x^2 and so this is going to start off $1 + x + x^2$. You just have one x^2 coming out of this. It's looking kind of boring here, but things get a little juicier.

How about x^3? Look, the third parentheses has an x^3 in it and it can multiply all the other ones. There's an x^3, but there's another way to get an x^3. When that x^2 in the second binomial multiplies the x in the first one, remember you add the exponents you've got x^2 times and x is x^3 there so there's going to be a second x^3 emerge. I'm going to write it this way. I'm going to show you where they're coming from so I'm going to say $+(x^3 + x^{2+1})$ showing where that other x^3 comes from. There are two x^3 and that shows the source.

How about x^4? That x^4 in the fourth parentheses can multiply all the other ones or you can get an $x^3(x)$ from the first and third parentheses to give you a second one. And so I'm going to write it $(x^4 + x^{3+1})$ showing you where that other x^4 came from. And one more, let me do x^5. This will I think establish the pattern. You can get an x^5 when the x^5 multiplies all the other ones; you get another one when the x^4 hits the x and you get yet another one when the x^3 hits the x^2. There's three x^5 and look at those exponents, $5, 4 + 1, 3 + 2$, we've seen those before. Those are exactly the decompositions of 5 into distinct pieces. Every such decomposition of 5 into distinct pieces will yield another x^5 and any x^5 that emerges here has to be built out of distinct pieces because look at $P(x)$ up at the top. All the expressions have different powers of x in it. How many x^8 are there going to be if you multiplied this out? You can look at how many ways to break 8 up into distinct pieces and there'll be six x^8 and you can multiply it and you'll see that's in fact the case.

What Euler has realized is that P(x), this thing he introduced up above is 1, it starts with a 1 and then we're going to say we're going to add up using the sigma notation, the sum is n goes from 1 to infinity of a certain number of x^n's. How many x^n's are you going to get? Exactly D(n), exactly as many ways of decomposing n into distinct pieces. When I write this expression what I'm saying is that P(x) is $1 + D(1)x^1 + D(2)(x^2) + D(3)(x^3)$, the number of ways, the number of x^n's you're going to get is precisely the number of ways of splitting n up into distinct pieces. That is evident from this P(x).

That's the first part of the proof. We'll store that. We'll get back to it. Next up we have to recall something called an infinite geometric series. If you've not seen it lately, here's what this is. It's the sum of $1 + a + a^2 + a^3$ forever and ever, another infinite series. This is called geometric, notice how each term arises from the preceding one by just raising the power one more, a, a^2, a^3 and you add these up. The question is what's the sum, what does this add up to, and it turns out to be $1/1 - a$. It's called an infinite geometric series. I don't have a whole lot of time to prove this, but let me just say here's a suggestion of why this works.

Suppose I take the $(1 - a)$ and multiply it by the series. You'd take $(1 - a)$ times the series $(1 + a + a^2 + a^3)$ and you multiply it out. When you hit the 1 times the 1 you get a 1, the $-a$ times the 1 gives you a $-a$. Then the 1 hits the a and the $-a$ hits the a so you get a $+a$ and a $-a^2$. You get a $+a^2$ and a $-a^3$ plus a^3 minus a^4, what happens is these terms come out with a plus minus plus minus plus pattern on them and if you look at this $1 - a + a - a^2 + a^2 - a^3 + a^3$, everything collapses except 1. If $(1 - a)$ times the series is 1, then the series is $1/1 - a$ and that's the result I'm going to need in a minute. There it is; there's the equation. Usually I should just say, we use this series, we use this formula from left to right, we got the series and we want to write it as a sum 1/1-a. Euler's actually going to use it from right to left as you'll see in a minute. So he's going to use it backwards sort of.

For the second main part of his argument he introduces Q(x), he calls it, which is $1/1 - x \times 1/1 - x^3 \times 1/1 - x^5 \times 1/1 - x^7$. Why is he introducing this?

Hang on there's a reason. You look at it you see the pattern. It's quite clear. We have $1/1 - x$ to the odd powers. Okay, so what's he want to do with this. The first thing he says is I don't want all these denominators here. Look at that first expression $1/1 - x$, look at the geometric series up above. I could replace the $1/1 - x$ by a series using the formula up above from right to left and replacing a and using x in its place, so $1/1 - x$ is just $1 + x + x^2 + x^3$, an infinite series. That is that first expression. The next one is $1/1 - x^3$. Look at the geometric series at the top. Instead of $1/1 - a$ I'm going to put an x^3 where the a is and just use the same formula again. Now I'm going to have $1 + a$, which is $x^3 + a^2$, which is $x^3(x^3)$ which is x^6 plus x^9 so when I take that second fraction $1/1 - x^3$ I replace it by this infinite series $1 + x^3 + x^6 + x^9$. You do the same thing with $1/1 - x^5$. The role of a is being played by x^5 so you get $1 + x^5 + x^{10} + x^{15}$, and so this is another expression for Q. It actually looks worse if you think about it. You have infinitely many infinite series here and you know what is Euler going to do with this.

He's got a little adjustment to make. You see that first expression $(1 + x + x^2 + x^3)$, he's going to write it this way $1 + x^1$ plus instead of x^2, x^{1+1}, sure that's x^2, instead of x^3, x^{1+1+1} and so on. For the second line $1 + x^3 + x^6 + x^9$ how about $1 + x^3 + x^{3+3}$ for 6, x^{3+3+3} for 9, the next one will be $1 + x^5 + x^{5+5} + x^{5+5+5}$. We're going to break these down and we're going to get the $Q(x)$ is equal to the product of these things $1 + x^1 + x^{1+1}$ and so on times $1 + x^3 + x^{3+3}$ and so on times $1 + x^5 + x^{5+5}$ and so on. That's $Q(x)$. What do you do with it? He says multiply it out infinitely many infinite series. Let's try it. The constant term is going to be $1 \times 1 \times 1$ from each of those parentheses and you're going to get a 1. How many x's are you going to get here? If you look at it, there's only one source of an x, that x^1 in the first parentheses hits all the other ones, no other way to get an x.

With x^2, that x^{1+1} in the first parentheses is an x^2; it can hit all the other ones. That's the only way to get an x^2. With x^3, there's two ways to get and x^3. The x^3 in the second parentheses can hit all the other ones or you go back to the first parentheses and there's an x^{1+1+1}, a second x^2. I'm going to write it $(x^3 + x^{1+1+1})$ to show you where the two x^3's come from. For x^4 you're going to get an $x^{3+1} + x^{1+1+1+1}$, two x^4. For x^5, think about it. There are three sources of an x^5. The x^5 in the third parentheses hits all the other ones, I can get an $x^3 \times$ an x^{1+1} or I can get from the first parentheses $x^{1+1+1+1+1}$, three x^5. Look at those exponents 5, 3, and 1, and 1, 1 and 1 and 1 and 1 and 1. Those are exactly the decompositions of 5 in to odd sum ends. You see they've got to be odd because look at how Q was written up at the top. All the exponents are odd.

What Euler sees is that $Q(x)$, this strange thing he introduced, is 1 plus the sum as n goes from 1 to infinity of a certain number of x^n, how many x^n's. How many x^n's are there? $O(n)$. Exactly as many x^n's is there are ways of breaking n up into odd sum ends. How many x^8 are there going to be here? The answer is 6 because there are 6 ways to break 8 up into odd pieces. That's the second part.

Here's the dramatic conclusion; remember what we showed. $P(x)$, which started off as $(1 + x)(1 + x^2)(1 + x^3)$. What Euler realized is 1 plus the sum of the $D(n)x^n$, so it's $1 + D(1)x^1 + D(2)x^2 + D(3)x^3$ and so on. The $Q(x)$, which started off as $1/1 - x \times 1/1 - x^3 \times 1/1 - x^5$ he saw it turned out to be 1 plus the sum of $O(n)x^n$. Remember this proof started so long ago I forget what we're trying to prove. We're trying to prove that $D(n)$ is always equal to $O(n)$.

If you look at this and you see how these spin out, we could prove that if only $P(x)$ and $Q(x)$ were the same thing. They're not, you can see, but if they were the same then their series would have to be the same, their expansions, and so the number $D(n)$, which is in front of x^n in the top would have to match $O(n)$, which is in front of x^n below all the way across the line. $D(1)$ would have to $O(1)$, $D(2)$ would have to be $O(2)$, $D(3)$ would have to be $O(3)$. You'd be done with your proof if only $P(x)$ and $Q(x)$ were the same. Then they'd have the same series, they'd have the same coefficients, proof over. If only they were the same.

But, look they're different; they're different. Euler says, or are they? You know what, they're really the same. They're the same thing and if I can show you that then I know that $D(n)$ and $O(n)$ are the same for all n and the proof is over. Here comes the dramatic conclusion. Why are these the same? They don't look the same to me. Here's how it goes, $P(x)$ remember is, when we first saw it, $(1 + x)(1 + x^2)(1 + x^3)$ forever and ever. I've written it there, but I've left some room between the terms. Euler draws a line; he's going to make a fraction out of this. Up in the numerator next to the $1 + x$ I'm going to put a $1 - x$. That's OK as long as I put a $1 - x$ below so I do. That keeps everything equal. Up next to the $1 + x^2$ I'm going to put a $1 - x^2$ and match it with a $1 - x^2$ below. A $1 - x^3$ on the top, a $1 - x^3$ on the bottom, a $1 - x^4$, $1 - x^5$, $1 - x^6$. All of these show up on the bottom to match their counterparts on the top.

That's still P(x), but now look at this, $(1 + x)(1 - x)$. I'm in the numerator, look at the first two terms, $(1 + x)(1 - x)$, that looks familiar, that's $1 - x^2$ and so those first two terms in the numerator take out the $1 - x^2$ term in the denominator. They cancel. The next two in the numerator $(1 + x^2)(1 - x^2)$, that's $1 - x^4$ cancel, $(1 + x^3)(1 - x^3)$ is $1 - x^6$, that takes the $1 - x^6$ in the bottom. You do this all down the line. What's left in the numerator? What's left is 1. What's left in the denominator? What's left are all the odds. You see them there. They didn't cancel. P(x) is just $1/(1 - x)(1 - x^3)(1 - x^5)$, but hey that's Q(x). P(x) and Q(x) are the same. Their series expansions are the same. Every one of their coefficients must be the same. D(n) must equal O(n) for all n, Q. E. D. Wow that is really a very nice argument. Nobody had ever gone down this path and he now starts the study of the partitioning of numbers. I think that's brilliant.

It was Frobenius, a 19th-century mathematician, who said this about Euler. He said, "Euler lacked only one thing to make him a *perfect* genius: He failed to be incomprehensible." I think you can read Euler's mathematics like this example, like the Basel Problem, give it a little attention, and you can follow it. You can understand it and you can realize that Euler was one of the greatest of all.

Gauss—The Prince of Mathematicians
Lecture 21

> Gauss kind of removed himself from the story [in his proofs], but on his behalf, people would say, "Well, wait a minute now; the architects of the great cathedrals don't leave the scaffolding up. You remove the scaffolding and you see the art—the cathedral behind." Gauss would remove all the extraneous material and leave just the gem of the theorem.

Carl Friedrich Gauss was born to an impoverished family in Brunswick, Germany, in 1777. As we saw with Euler, Gauss's mathematical gifts were obvious from a young age. In one famous story from his childhood, Gauss correctly added the first 100 whole numbers in a matter of seconds. He saw that adding $1 + 100 = 101$, $2 + 99 = 101$, $3 + 98 = 101$, and so on. He multiplied 100 by 101, then divided the product to get 5,050. Obviously, he was no ordinary student.

By the age of 15, Gauss's academic training was being funded by the duke of Brunswick. The young Gauss read the works of Newton, Euler, and others and became inspired by the great predecessors who had blazed the path for him. As an adolescent, he kept a notebook with his mathematical discoveries and conjectures; this is now a treasure-trove of great results, anticipating mathematics that would come forth in the 19th century. Gauss's first great discovery occurred in 1796, when he expanded on the idea that various regular polygons can be constructed geometrically. In 1799, he obtained his doctorate; his dissertation was the first proof of the fundamental theorem of algebra: Any real polynomial can be factored into the product of real linear and/or real quadratic factors. In 1801, he published his *Disquisitiones arithmeticae*, a deep and difficult book on number theory. Also in 1801, Gauss became famous for finding the "missing" asteroid Ceres using mathematics.

In 1807, Gauss became the director of the observatory at Göttingen in Germany and remained there for the rest of his career. He continued to make many contributions to mathematics and the sciences, including the theory of least squares, which underpins the subject of regression, and the

Gaussian, or normal, distribution in probability, which permeates all of statistical analysis.

As mentioned earlier, Gauss expanded the idea that regular polygons could be constructed with a compass and straightedge. In particular, he dealt with the heptadecagon, the regular 17-sided polygon. Earlier, Descartes had shown that it was possible to construct geometrically any length that is built from whole numbers and the operations of addition, subtraction, multiplication, division, and extraction of square roots. Expressions built in this way are now called quadratic surds (surds means "roots"). Gauss asserted that he could construct the 17-gon from the construction of $cos2\pi/17$. Using something called the 16th-degree cyclotomic polynomial, Gauss concluded that $cos2\pi/17$ could be written as a quadratic surd, which means that it is constructible. The generic statement of Gauss's theorem here is as follows: If $P = (2^2)^n + 1$ is prime, then a regular P-gon can be constructed with a compass and straightedge. Gauss's insights into such constructions and many other areas of mathematics opened doors for explorations that were fundamental to abstract algebra in the 19th century. ∎

> **Gauss expanded the idea that regular polygons could be constructed with a compass and straightedge.**

Suggested Reading

Dunnington, *Carl Friedrich Gauss*.

Ore, *Number Theory and Its History*.

Tent, *The Prince of Mathematics*.

Trudeau, *The Non-Euclidean Revolution*.

Questions to Consider

1. Use the trick attributed to little Carl Friedrich Gauss to find the sum of the first *thousand* whole numbers. Without the trick, this is hideous; with it, it's a snap.

2. Here's how to construct the magnitude $\sqrt{5}$ with a compass and straightedge (indeed, any square root can be done in analogous fashion):

First, draw line AB of length 5 and extend it with segment BC of length 1. Thus, AC is of length $5 + 1 = 6$ units. Bisect AC at O, and draw a semicircle with center O and with AC as the diameter, as shown below. From B, construct BD perpendicular to the diameter, with point D on the semicircle. All of these constructions can be accomplished with compass and straightedge.

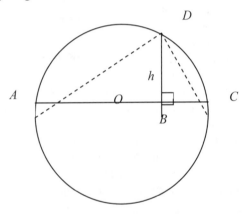

Now, $\angle ADC$ is a right angle (why?), and thus, $\triangle ABD$ is similar to $\triangle DBC$ (see Lecture 4). Use the proportional sides of these similar triangles to show that h, the length of BD, is exactly $\sqrt{5}$, as required. Thus, if we can construct a magnitude with compass and straightedge, we can construct its square root as well.

Gauss—The Prince of Mathematicians
Lecture 21—Transcript

In this lecture, we'll meet Carl Friedrich Gauss. Gauss was six years old when Euler died. It's easy to see this as a passing of the baton from one towering figure to another. In my book, the mathematical Mount Rushmore contains Archimedes, from classical times, the great Newton, Euler, and Gauss. Our object in this lecture is to meet Mr. Gauss.

This is a picture of him as a young man; he's very good looking. Notice that he's not wearing the head warmer because he didn't do the wig. This is a more modern era. He looks almost like he's wearing clothes we would recognize maybe from a Jane Austen novel. He lived about that time. He was born to a poor family in Brunswick, Germany; he did not have the advantages of wealth. It was said that little Gauss could calculate before he could speak. He was obviously gifted in terms of his work in mathematics. There's a story that at the age of three he was correcting errors in his father's account books. This might be legendary, but you can sort of imagine this little boy, barely able to see over the table, but pointing out that this column of numbers had been added up incorrectly.

There's a famous story from Gauss' childhood. It involves something that happened at his school where he was asked to sum the first hundred whole numbers. This story's authenticity might be suspect, but the teacher wanted a break and so asked the students to add up the first hundred numbers figuring that would occupy them for some time. In an instant, little Gauss comes up and puts his slate on the desk. It's got the correct answer. How in the world did he do this? Did he add up a hundred numbers in a split second? No, he saw something.

The way the story goes is this. You imagine the sum of the first hundred numbers, $1 + 2 + 3$, all the way up to 100 and call that S for sum. Little Gauss imagined writing that same sum right below, also S. But, this time, write the numbers in descending order. It's the same sum, but now I'm going to take $100 + 99 + 98$ down to 3, 2, 1. You haven't changed the value at all, just the order in which you've written them. Now draw a line and add in columns. On the left side of the equals sign, you get $2S + S$ equals 2S obviously. That's easy enough. Look on the right side. That first column you have a $1 + 100$, that's 101. Next to it is $2 + 99$, well that's 101, $3 + 98$, again 101. Each time we raise the number above, we lower the number below, same sum 101. It continues all the way across $98 + 3$, 101, $99 + 2$, 101, $100 + 1$, 101, so 2S is the sum of all these 101's. How many are there? There are exactly as many as there are columns and there's 100 columns. There's 100, 101's, which even I could do in my head is 10,100. If 2S is 10,100, S is half that 5,050. In other words, the sum of the first 100 numbers is 5,050 and in a shorter time period than it took me to explain it little Gauss had seen it, written on his slate and handed it in. Obviously this was not an ordinary student here.

By the age of 15, Gauss's academic training was being funded by the Duke of Brunswick. His talents were so evident that the Duke said we have to support this young man through his training even though he is not a person of means, we'll find the money. Gauss certainly benefitted from the training he got. He read the works of Newton, Euler, and others, and became inspired by the great predecessors who had blazed the path for him. He said, " I became animated with fresh ardor, and by treading in their footsteps, I felt fortified in my resolution to push forward the boundaries…" He read the masters, and now he's ready to push further.

As an adolescent, he kept a notebook with his mathematical discoveries and conjectures, written in a very cursory fashion, but he would just make these little notes of this or that that he discovered or believed. This is now a treasure-trove of great results as he's anticipating so much mathematics that would come forth in the 19th century. His first great discovery occurred in 1796 and he announced it this way. Gauss wrote, "It is known to any beginner in geometry that various regular polygons, for instance the triangle, the square, the pentagon, the 15-gon, and those which arise by the continued doubling of the number of sides of one of them are geometrically constructible." What he's saying there is that people know that you can take out a compass and straightedge and do an equilateral triangle, a regular 3-gon, of course that was the first thing Euclid did in Book 1. You can do a square, the regular 4-gon, a regular pentagon, a regular 15-gon. We've seen all this and Gauss acknowledges, "One was already that far in the time of Euclid."

That was known. He writes, "so much the more, methinks, does my discovery deserve attention ... that besides those regular polygons a number of others, for instance, the 17-gon, allow of a geometrical construction." What he's announcing here is that you can take out your compass and straightedge and construct a regular 17-sided polygon, which no one had ever even imagined was possible. His teacher appended a little note to this announcement and said, "It is worth mentioning that Mr. Gauss is now in his 18th year and has devoted himself here in Brunswick with equal success to philosophy and classical literature as well as higher mathematics." This is an endorsement from the teacher. Gauss later attributed his mathematical career to this discovery of the regular 17-gon. That convinced him that maybe he had a future in mathematics. Otherwise he might've become a philosopher or a classicist.

In 1799, he obtained his doctorate from the University of Helmstadt. His doctoral thesis, his dissertation was the first proof of the fundamental theorem of algebra. What's that? It's this statement that, "Any real polynomial can be factored into the product of real linear and/or real quadratic factors." You give me a real polynomial maybe of degree 80 or 800, you can shatter it down into the product of first and second degree pieces. It's called the fundamental theorem of algebra.

For example, if I looked at $x^4 - 1$, that's a fourth degree polynomial, you can obviously break that into $(x^2 + 1)(x^2 - 1)$ and further break the $(x^2 - 1)$ into an $(x + 1)(x - 1)$ and so the fourth degree is written as the product of a quadratic, the $x^2 + 1$ and the two linear pieces. That's the phenomenon that he's getting at in his dissertation, but he proves this is true of any real polynomial. It's an existence proof. He shows that there exists such a factorization; it doesn't demonstrate how to find it explicitly. If I gave you an 80th degree polynomial it wouldn't show you what the factors were, but it shows they exist. This was a giant advance. Euler in the previous century had given a clever, but ultimately unsuccessful proof of the fundamental theorem of algebra in 1749. He had advanced that he thought it was true. He gave a proof and it wasn't a valid proof. Euler had sort of flubbed it.

Gauss in his dissertation mentioned Euler's proof and criticized Euler saying that Euler's argument was "shadowy," and so Gauss then proceeded to prove the result. He was willing to take on even Euler in his dissertation. The fundamental theorem of algebra as the name suggests is really important and how many people prove a fundamental theorem in their dissertation? Not many people do. In fact, it's been said that Gauss' was probably the greatest doctoral dissertation in the history of mathematics. It proved the fundamental theorem of algebra.

In 1801, he published his *Disquisitiones arithmeticae*. This is a deep and difficult book on number theory and here we are again with that old subject, number theory. Euclid, Fermat, Euler, and now Gauss puts his name in the chain of great number theorists. The *Disquisitiones arithmeticae*, but it's very hard going. It's very difficult reading. Someone has said that the book is no easier to understand than its English translation than it was in his Latin. It doesn't matter if you translate it, it's still going to be hard to get through.

This was sort of typical of Gauss's writing. He was terse. He left out a lot. This contrasts with Euler. Euler's writing was very discursive, very expository. He would tell you what he's thinking. He said I would try this and that didn't work very well now I'll try that. As you read it you're following his thought processes right along. It's very user-friendly shall we say. Not so with Gauss—Gauss would remove the extraneous details. He wasn't going to tell you what he was thinking. He would just show you the steps and sometimes only every other step and you kind of had to fill in the gaps yourself. So reading a Gaussian work is a lot harder than reading a Eulerian one. Gauss has been compared to the Fox that walks through the sand dragging its tail behind it so as to erase its tracks so you don't know that it even passed by. There's no record of the fox having come by.

Gauss kind of removed himself from the story, but on his behalf people would say well wait a minute now the architects of the great cathedrals don't leave the scaffolding up. You remove the scaffolding and you see the art, the cathedral behind. Gauss would remove all the extraneous material and leave just the gem of the theorem. There's a contrast in the Gaussian and the Eulerian way of writing mathematics.

In 1801, Gauss became famous for finding a missing asteroid. They lost an asteroid and he found it. That requires a little story too here. What happened was that the asteroid Ceres had been observed by Giuseppi Piazzi in January of 1801. Piazzi discovered it, he'd had a few readings of where it was, but then it was lost in the glare of the sun. The asteroid was too near the sun. You couldn't observe it anymore. Now what happens is that you know that it's going to come out on the other side of the sun eventually and then you could look for it. But, they couldn't find it. They didn't have enough information of its orbit and so although they sort of suspected it was out there in the sky some place, they couldn't find Ceres.

Gauss is given the limited data that Piazzi had acquired. He had a little bit of information of where this asteroid was and what he had to do is try to use this to figure out its orbit and tell the astronomers where to look for this thing. Gauss used a find theory of celestial mechanics that he had worked on and a personal invention of what we now call the method of least squares, it's quite important today. Using this he made this prediction and he told the astronomers where to look for Ceres and they turned their telescopes to that spot in the sky and voila there it was. He found the lost asteroid using his wonderful mathematics, his fabulous mind.

In 1807, he became the director of the observatory at Göttingen in Germany and remained there for the rest of his career. He was an astronomer as well as a mathematician and he made other contributions to the sciences. As I mentioned the theory of least squares has become a very important subject, it now underpins the whole theory of regression and regression is something that's used in business and economics and psychology and all sorts of subjects. If you dig deeply backwards you'll find the theory of least squares there. That's one of Gauss' great contributions.

He worked on Geodesy, the study of the earth's surface, he did maps, measured the earth. He worked in differential geometry, a very abstract theory of surfaces in space. It's a very hot item right now in mathematics that tends to be traced back to Gauss. Along with Wilhelm Weber, he worked on the theory of magnetism. Gauss and Weber jointly did experiments, studied magnetism, they kind of invented the telegraph although his name isn't attached to that, but we do have a unit of magnetic flux called the Gauss and people honor him with that unit.

He also looked at something called the Gaussian distribution in probability. We call it the normal distribution, the bell curve. That's pretty important; it permeates all of statistical analysis. Gauss was one of the first to look at this carefully to understand that this shows the distribution of error, random error follows this bell curve, this Gaussian. When the Germans still issued marks rather than the euro, this was their 10-mark note and here's Gauss on the bill, you can see him. If you look just to the left of his cheek and a little above that 10 you see a bell curve, the Gaussian distribution.

In Europe it's more likely to be called the Gaussian, here in the U.S. I think it's more likely to be called the normal, but it's the same distribution and it's just as important no matter where you are. Gauss also worked in non-Euclidean geometry. I'm going to talk more about this in the next lecture, but let me just say that what was going on was he was going to replace Euclid's theory of parallels with one in which triangles have fewer than 180 degrees in their angle sum. With Euclid's geometry if you take his 5th postulate you prove that a triangle has 180 degrees. He said two right angles in its angle sum. But, Gauss fiddled with that postulate and came out with triangles that have less than 180 degrees and studied this, what we now call, non-Euclidean geometry. He said it's "… a special geometry, quite different from ours, which is absolutely consistent and which I have developed quite satisfactorily …" He believed he had followed the path into non-Euclidean geometry and he understood it. As I say, we'll talk more about this in the next lecture.

On the personal side, he married wife Johanna, and he had two children. Then, she died tragically, so he remarried wife Minna and there were three more Gaussian offspring. He was known as "… glacially cold…" as a personality and indeed in the family there weren't warm relationships; it wasn't a warm and fuzzy family. There were falling out scenarios among the children and the parents. Gauss was conservative in his mathematics, conservative in his politics. He had a motto, pauca sed matura, which is Latin for few but ripe. The idea was he would publish few results. He wasn't like Euler that filled 75 volumes and 25,000 pages. Gauss's collected works were much more modest so they're few, but boy are they ripe. He's not going to publish them until they're just perfect and ready to be plucked and he gave the world some very ripe mathematics.

I want to tell you a little bit about the regular heptadecagon as it's called, the regular 17-sided polygon, hepta, seven, deca, ten, so it was that Gauss pursued this and showed that you could construct such a thing with a compass and straightedge much to the surprise of everybody—this result that all prior geometers had missed even though it was lying right under their noses.

A couple of preliminaries you need, one is from Descartes. René Descartes had shown in the 17th century that it is possible to construct with a compass and straightedge any length that is built from the whole numbers and the operations of addition, subtraction, multiplication, division, and the extraction of square roots. If you can build an expression from the whole numbers and these algebraic operations up to and including square rooting things then you can actually construct that length. These are now called the quadratic surds, S-U-R-D-S, which means roots. It's actually a contamination of a word from Al-Khwārizmī, our old friend from medieval Islam, the great algebraist, the word surd comes from him. A quadratic surd thus would be something like the $\sqrt{2}$ because that's made out of the number 2 square rooted.

You can construct a length of the square root of 2 with a compass and straightedge or how about $3 + \sqrt{5}$, yeah, because it's 3 and 5 $\sqrt{}$ add, all of that's legit, you can construct that or even some monster like this $1 + 2\sqrt{3} + 4\sqrt{5/6} - \sqrt{7/8}$. That looks pretty nasty, but it's made out of the whole numbers added, subtracted, multiplied, divided, and square rooted. Maybe square roots within square roots, but that could be constructed with a compass and straightedge. Any quadratic surd could be so constructed, that from Descartes.

Second preliminary Gauss needed was this fact. You can construct a regular heptadecagon, a regular 17-gon if you can construct the $\cos 2\pi/17$. He's sort of replacing the challenge and instead of having to construct the 17-gon he said look if you just can construct the $\cos 2\pi/17$ that'll be enough and I'll show you then how to go from there to the 17-gon. This takes a little bit of background. First of all, what's this $2\pi/17$ doing here? Remember in radians 2π is one loop around the circle. An angle with measure $2\pi/17$ is exactly 1/17 of the way around. But, if I can do that I'm sort of onto the problem of going exactly 1/17 of the way around, but exactly 1/17 of the way around is what you need to do a regular 17-gon. Somehow there's this connection and here's how it goes. Let's suppose I have constructed the $\cos 2\pi/17$, so somehow I got out my compass and straightedge. I did all my work and I did it and there it is. It's red. I've constructed that length.

How do I go from that and make a regular 17-gon? Here's what Gauss says you do. You get a point O to be the center of a circle, draw a circle of radius 1. There's an arc, I just have the first quadrant there and take your little red line, your $\cos 2\pi/17$, which you have constructed and copy it along the horizontal axis from O to A, so that length from O to A is $\cos 2\pi/17$ and the length you have constructed. From A you put a perpendicular upward until it hits the circle at B and you draw OB, the radius, which of course is also 1 because this is the unit circle and this forms a triangle here, a right triangle, OAB.

Let me call the angle down there $\triangle AOB$ theta; we'll think of that in radians. Now we want to look at $\triangle AOB$ and get the cos in the picture. We've talked about sines in here. I don't think we've previously mentioned cos, so let's just remember that the cos of the angle theta is the adjacent side over the hypotenuse. Remember the sin is the opposite over the hypotenuse, cos is adjacent over hypotenuse. In $\triangle AOB$ the adjacent side is OA, the hypotenuse is OB and so my ratio of OA to OB, remember OA was the cos of $2\pi/17$, OB was the radius of 1, so the cos of theta is just the cos of $2\pi/17$.

That means if you look at the drawing there, if the cos of theta is the $\cos(2\pi/177)$ since we're in the first quadrant, you can conclude that theta is $2\pi/17$. That angle there, $\triangle AOB$, which I had called theta, I'm going to remove the theta and replace it with $2\pi/17$. That's what that must be. But, that means that the arc thus created, the arc from B to C is exactly $1/17$ of the way around. Notice I haven't drawn this to scale. That picture doesn't look like $1/17$, but I wanted it to be big enough to see what was going on. In any event, the arc BC is $1/17$ of the way around so if you draw the line along BC, the cord from B to C that is the side of the regular 17-gon, get out your compass copy it 17 times—bingo, you've got it. That's the second fact. If you can construct the $\cos 2\pi/17$, you can do a regular 17-gon.

The third and really most amazing part of Gauss's reasoning is this. He looked at what's called the 16th degree cyclotomic polynomial $x^{16} + x^{15} + x^{14}$ all the way down to $x^2 + x + 1$. He figured out the roots of this and using that in some really amazing and ingenious mathematics he concluded that the cos$2\pi/17$ could be written as $-1/16 + 1/16\sqrt{17} + 1/16\sqrt{34 - 2\sqrt{17}}$, but there's more. There's another gigantic square root with a $1/8$ in front of it and I'm not even going to read this last part, but notice there're square roots within square roots within square roots. This is exactly what the cos$2\pi/17$ is. Nobody, nobody had ever seen anything like this. But, Gauss proved it by studying that 16th degree polynomial out there.

This is the hard part of the proof. This would take many, many, many lectures to go through so I'm just going to tell you he did this, but here's the critical issue. Look at that monstrosity there; it is a quadratic surd. It's nothing but the whole numbers added, subtracted, multiplied, divided, and square rooted, so that's a surd. Although some people when they look at is say that's AB-surd. That's really ridiculous, but this is a surd.

Here comes the argument that Gauss cobbles together. He said look I just showed you that the cos$2\pi/17$ is a quadratic surd because I figured out what it is and it had nothing by square roots and numbers in it. The cos$(2\pi/17)$ is a quadratic surd, but that means it's constructible from what Descartes had said. You can construct quadratic surds so with a compass and straightedge I can actually make a length cos$(2\pi/17)$, but then Gauss had shown that if you could do the cos$(2\pi/17)$ you could do the regular 17-gon in the fashion I showed you.

What's the conclusion? A regular 17-sided polygon is constructible with compass and straightedge. It's astonishing and it was from an 18-year-old kid. Actually, Gauss proved more than this, more than just the 17-gon. He proved this amazing fact that if the number $P = 2^2$ to the power n, so we have a stack of exponents, 2^2 to the n plus 1. If that number is prime, 2^2 to the n plus 1 then you can construct a regular p-gon with a compass and straightedge. This is his generic theorem.

Let me just look at a few values for n here and see what's up. If n is 0 then the number is 2^2 to the 0 plus 1. What you do is you go up to that exponent, you look at the 2^0 up there, which is 1 and so I have 2^1 plus 1, which is 3, 2 and 1 is 3. That's a prime. Gauss's theory says you can construct a regular 3-gon, an equilateral triangle. People had known that before Gauss's theory, but it's nice to see that falling out of his ideas. If n is 1, 2^2 to the 1 plus 1, now it looks like 2^2 plus 1, 2^1 up there is a 2, 2^2 plus 1 is 5, that's a prime. His theory says you can do a regular pentagon. Again we knew this from Euclid, but there it comes from his theory, but now he moves into new territory. If n is 2, I'm looking at 2^2 to the 2 plus 1. That 2^2 up there is a 4, 2^4 plus 1 is 17, and there's the regular 17-gon that he discovered.

If n is 3, I now have 2^2 to the 3 plus 1, 2^8 plus 1 is 257, that's a prime and so you can construct a regular 257-gon by his theory. If n is 4 you have 2^2 to the 4 plus 1, which is, these numbers are getting huge, 2^{16} plus 1 is a 65,537-gon, which is a prime and that could be constructed. There's not much point to it because if you actually constructed a regular 65,537-gon it would look like that. Can you can imagine how big the radius would have to be before you could see the little sides there, a little like a circle, but it could be done. That's the amazing thing. It can be done. It falls out of his theory.

The next candidate would be 2^2 to the 5 plus 1, which is 2^{32} plus 1. If you multiply that out, you get 4,294,967,297. Can you do one of those? That doesn't come out of his theory because, guess what, that's not a prime. That number isn't a prime. It factors into $641 \times 6,700,417$ and the person that first saw that was Euler who was looking at this number for a completely different reason and managed to factor it. Since that's not a prime it doesn't fit Gauss's pattern, nor have any of the other numbers yet explored 2^2 to the $n + 1$ turned out to be prime. It was just those first few at least as far as we know at the moment where his theory works.

I should stress that the importance of all this is not because we really care about constructing regular polygons with compasses and straightedges. We don't have to do this for any particular reason, it's not that this has any practical use, but rather it's theoretically so fascinating that it can be done and Gauss' insights open doors to explorations into abstract algebra that became so fundamental as we move into the 19th century. You can trace a lot of the theory of abstract algebra back to things like this from Gauss.

Here's a picture of Gauss late in life and his epithet by then had become the *Princeps mathematicorum*, the Prince of Mathematicians. That was certainly accurate.

The 19th Century—Rigor and Liberation
Lecture 22

What mathematicians decided was if you studied Euclid's geometry, where triangles all have 180 degrees, great. If you studied this non-Euclidean geometry, where they don't, great. ... [B]ut certainly, they can't both be true. ... Logically, [however,] these are sound, these are consistent, these are legitimate pursuits, and thus, mathematics is freed from the constraints of physical reality by the kinds of pursuits that Gauss, Bolyai, and Lobachevski began.

As we saw with the heroic 17th century, the 19th century was also a significant period in the history of mathematics. The achievements of this century can be broken into four themes: rigorization, liberation, diversification, and abstraction. We'll look at the first three of these in this lecture and the fourth in Lecture 23.

We begin with rigorization as it relates to the underlying ideas of calculus, particularly the notion of the derivative, which is the slope of the tangent line to a curve at a point. It shows how steeply the curve is rising or falling at that point. The slope of a line is the rise over the run, but we have only one point where the tangent occurs; thus, we usually approximate the slope by constructing a right triangle on the curve. But what if we want the exact slope of the tangent line? Newton used what he called "vanishing quantities"; that is, as the ratio of the change in y to the change in x ($\Delta y/\Delta x$) grows smaller and smaller, the slope is that ratio at the moment those values reach zero . Many found this definition a little hazy, as they did Leibniz's idea of infinitely small quantities. The process of rigorization

The French mathematician Cauchy defined the term "limit."

to repair the foundations of calculus occurred in the 19th century, primarily through the efforts of the French mathematician Augustin-Louis Cauchy. Cauchy based the derivative on limits rather than vanishing or infinitely small quantities. For Cauchy, the derivative is the limit as Δx of the ratio $\Delta y/\Delta x$ goes to zero. Cauchy provided a definition of the term "limit" that was later refined by Karl Weierstrass.

> **One of the surprising results of non-Euclidean geometry is that different triangles have different angle sums; there is no single value for the sum of the angles of a triangle.**

Also during this period, mathematics was freed from the bounds of reality. This liberation emerged in the study of non-Euclidean geometry. Recall that Gauss had explored triangles whose angles sum to less than 180 degrees. Two other mathematicians, the Hungarian Johann Bolyai and the Russian Nikolai Lobachevski, invented similar geometries at around the same time. One of the surprising results of non-Euclidean geometry is that different triangles have different angle sums; there is no single value for the sum of the angles of a triangle.

Finally, the third trend in the 19th century is diversification with regard to the mathematical players. Until this time, women were not supposed to pursue mathematics because the discipline was not seen as feminine. Further, formal education was denied to most women. One of the women who began to break down these barriers was Sophie Germain, a self-taught French mathematician. As a child, she had to hide her mathematical studies from her parents, and she was not allowed to attend university classes. Nonetheless, she carried out significant work in applied mathematics and number theory. Her most well known contribution is the Germain prime: A number p is a Germain prime if $2p + 1$ is also prime. Other women who followed Germain in the 19th century include Sophia Kovalevskaya of Russia and Grace Chisholm Young of Britain.

In our next lecture, we'll look at the fourth great trend of the 19th century, abstraction, as it appears in the set theory of Georg Cantor. ∎

Suggested Reading

Boyer, *The Concepts of the Calculus*.

Child, ed., *The Early Mathematical Manuscripts of Leibniz*.

Dunham, *The Calculus Gallery*.

Edwards, *The Historical Development of the Calculus*.

Grabiner, *The Origins of Cauchy's Rigorous Calculus*.

Grattan-Guinness, *From the Calculus to Set Theory*.

Osen, *Women in Mathematics*.

Trudeau, *The Non-Euclidean Revolution*.

Question to Consider

1. Those interested in the history of calculus might consult my article "Touring the Calculus Gallery," which can be found online at http://mathdl.maa.org/images/upload_library/22/Ford/dunham1.pdf. This describes in more detail the work of Cauchy, Weierstrass, and their colleagues who shored up the logical foundations of the calculus.

The 19th Century—Rigor and Liberation
Lecture 22—Transcript

A few lectures ago, I talked about the 17th century, the so-called heroic century in the history of mathematics. It was the century that brought us logarithms, analytic geometry, and at century's end, the calculus itself. As centuries go, the 17th set a very high bar. But, the 19th century is not far behind. It too was a very significant hundred-year period in the history of mathematics. It is the 19th century I want to talk about in this lecture.

Of course, any century that begins with Carl Friedrich Gauss on the stage is going to be a pretty important one. It's off to a great start. Gauss was 23 in 1800 and of course, the bulk of his career was spent in the 19th century. But, he wasn't the only actor, he was not the only contributor, and the achievements of the century go well beyond those of Carl Friedrich Gauss.

What I would suggest is that we can break the century's achievements into four themes, if you will, four trends that seem to be addressed with great success in the 19th century. Let me tell you what those four are. First of all, there is what I will call rigorization. It's a strange word. What this is about is the process by which the foundations of mathematical ideas are made more rigorous, more precise, and more logically exact than they had been in the past. This is particularly critical when it comes to the foundational ideas of the calculus. Newton and Leibniz had invented the calculus and the Bernoulli's and Euler had refined it, but there were still problems with the foundations. These were evident and they needed to be fixed as we move to a more rigorous calculus. That was a great achievement of the 19th century.

Second, I would say, was the movement toward liberation, by which I mean that mathematics no longer was tied into physical reality. It was freed from the constraints that were imposed upon it, say, by the physicists. Previously, mathematicians would study the mathematics that related directly to the real world; that seemed to be the mission. But, by the end of the 19th century, that no longer was the case. Mathematicians could study logical structures, logical systems that didn't necessarily have a connection to reality. They were freed from this. The event that led to this liberation was the appearance of non-Euclidean geometry in the 19th century. I want to talk about that.

Third there is what I would call diversification. By this I do not mean diversification of topics, but diversification of the cast members, the players, who was allowed to do mathematics and in particular I'm talking about the appearance of women upon the scene. If you think about it all the faces we've seen thus far have been male. All the mathematicians we've encountered have been men, women have to break through some pretty serious barriers and that happens in the 19th century, so we'll meet a few of those heroines who broke through. Finally there was the move toward abstraction and generalization. This was a major trend in the 19th century, best typified by the work of Georg Cantor, his abstract set theory and I will talk about Cantor in the next lecture. For this lecture, let's just focus on those first three—rigorization, liberation, diversification, with abstraction to come next.

Rigorization, what's this about? This was with regard to the underlying ideas of calculus and, in particular let's look at the notion of the derivative. This is one of the key ideas in calculus. This is the slope of the tangent line to a curve at a point. There's a curve and there's a point on it, and so what I want to do is draw the tangent line there so I have it as a dotted line. What is of interest, the derivative is the slope of that line. What it's showing you is how steeply the curve is rising at that point or falling if it's going downhill.

Imagine this. Imagine you're running along the curve, you're moving along that curve and when you get to the point that I've shown there you suddenly just shoot off into the straight direction and which you're headed at that instant. That would be the direction of the tangent. The slope of the tangent is showing you the direction you're moving along a curved object, along a wavy line for instance at a point. It's a very sophisticated idea and very important. It's called the derivative.

How do we do this? How do you find it? You see the trouble is if you want the slope of a line you need to points on it. The slope is the rise over the run and I only have one point where the tangent occurs there so you've got to use some ingenuity to figure out what this slope is. What is normally done is the following. They make a little triangle here. You go over a little distance called Δx and up a little distance called Δy. As the picture shows Δx is the horizontal distance of course of the little right triangle, Δy is the vertical and I go up to the curve. Then I draw the line that completes the hypotenuse.

The argument that I can't necessarily find the slope of the tangent line, but I can find the slope of that little blue line I just drew, which is going to be an approximation. The slope of the tangent at the point is approximately the slope of the blue line, which is the rise over the run as always, which means the change in y over the change in x, which in my picture is clearly Δy over Δx. You think of Δy over Δx as an approximate value for the slope of the tangent so that much is easy, but then what? What do you do to get the exact slope of the tangent line—not just an approximation, but the exact one? This is where things got a little messy.

Newton, when he cooked up his calculus, his fluxions used vanishing quantities to get at the exact slope of the dotted line, vanishing quantities. Here's the picture and let me quote him. He said, "By the ultimate ratio of vanishing quantities," and what he called the ultimate ratio is what we're looking for, the derivative, "is meant the ratio of the quantities not before they vanish, or afterwards, but with which they vanish." What's he talking about? We have this $\Delta y/\Delta x$, so you have this ratio. He imagines that these vanish, they head towards zero, so the Δx is getting smaller, the Δy is getting smaller, the little blue line there would therefore be getting ever more close in its slope to the dotted line, and you'd be heading towards something that you want. He says as these get smaller and smaller at the instant they vanish that ratio is the ultimate ratio. That's the derivative.

I would agree with him that you don't want the ratio before they vanish which is what he says. When he says you don't want to take the ratio after they vanish, I don't even know what that means, after they vanish, you know, the question is how long after they vanish are you going to come back and look at these things. They're gone. No, no, he wants the ratio with which they vanish. You imagine here they go and poof at that instant they vanish you try to capture the ratio. You can see that's not really a satisfactory definition upon which to build your calculus.

That looks a little hazy and it's not just I who thought that; Bishop George Berkeley certainly thought that back in the 18th century. Berkeley was a cleric; he was a philosopher; he was a mathematician; and he was unhappy with Newton's vanishing quantities. He said these are "... neither finite quantities, nor quantities infinitely small, nor yet nothing." Famously he said, "May we not call them the ghosts of departed quantities?" This is a pretty caustic rebuke, as these things disappear as the Δx goes to zero, the instant they vanish you look at their ghosts and take that ratio. You can see Berkeley saying the foundational issue is not set here. This is trouble we've got to repair this.

Leibniz when he invented his calculus had a slightly different attack on this. If you look at the picture, you have the little triangle, but now he thought that the horizontal displacement was dx, he called it, the vertical displacement dy where these were differentials, these were infinitely small quantities, so they didn't have to vanish, they were already infinitely small, but what's that mean. Does that mean zero? If something's infinitely small it's zero. It didn't seem to be quite that, but if it's not zero can't you always make a non-zero quantity even smaller. How can you be infinitely small? Just take half of it; wouldn't that be smaller still?

The logic there, the basic philosophical meaning of infinitely small, isn't very clear either and Bishop Berkeley was on the scene criticizing Leibniz as well. He said, "Infinitely small quantities present "an infinite difficulty" to understanding. Berkeley wasn't buying either of these approaches to try to explain the foundations of calculus. He said, "The further the mind analyseth and pursueth these fugitive ideas, the more it is lost and bewildered." He's suggesting we need to work on this.

He didn't deny the outcome—the derivative of x^2 is $2x$. He understood that's true. It was the logic behind it that he found unsatisfactory. Berkeley said, "Error may bring forth truth, though it cannot bring forth science." Maybe the results came out right, but they were based on error, maybe even compensating errors. It was not science; it needed the attention of the mathematical community to rigorize this great subject. That process of rigorization to repair the crisis and the foundations of calculus happens in the 19th century when mathematicians try to make calculus every bit as precise and rigorous as say Euclid's geometry was. It was a great adventure and it was not simple, but it finally came to pass in the 19th century and the name associated with the rigorization of the calculus is Augustin-Louis Cauchy, a French mathematician who was Gauss's almost exact contemporary and had

Gauss not lived Cauchy would've been the Prince of Mathematicians. He was an extraordinarily talented person.

Here's a portrait of him from his youth and I have something new here, a photograph. We actually now can see what these people looked like in a photographic record. This is Cauchy in later years. What Cauchy said is we're not going to build the theory of calculus, the derivative on the basis of vanishing quantities as Newton had nor infinitely small quantities as Leibniz had, but Cauchy says the trick is to do this with limits. He built his calculus upon this bedrock of limits. Whenever he looked at $\Delta y/\Delta x$ for instance, that ratio of the height and base of the little triangle, then to get the exact slope of the tangent he would take the limit as Δx goes to zero of that $\Delta y/\Delta x$ and that's the derivative for Cauchy. If you've had calculus you know this is how it's done today. You do limits to define derivatives—why Cauchy? He gave us this great idea.

There is a problem here though. Just to say the word limit doesn't really help. What is a limit? I need to define it. If this is actually going to work, it's not just we have a term on the board, but we have to give it meaning and here was Cauchy's 1821 definition of limit. He said, "When the values successively attributed to a variable approach indefinitely to a fixed value, in a manner so as to end by differing from it by as little as one wishes, this last is called the *limit* of all the others."

There is his definition. You have a fixed quantity, your variable is approaching it, so as to end by differing as little as you want from that limit. If you want to get within $1/10^{th}$, you get within $1/10^{th}$. If you want to get within $1/1000^{th}$, you'll get within $1/1000^{th}$. You can get as close as you want. That becomes his characteristic property of limits. Then using this definition he defines derivatives and continuity and integrals and all of calculus comes from this and he does it in a very precise and methodical fashion. It's a great, great achievement. However, it's still not quite the last word. If you look at the definition that Cauchy provided, it's very wordy, it doesn't look quite as mathematical, it's almost more literary. We need to sort of clean this up and get the symbols on the page, get rid of all these excess words.

That was done by Karl Weierstrass. Weierstrass a German mathematician active in the latter part of the 19th century. He and his followers put forth this definition of limit, which is now the one we use. Here it comes. I'll show this to you, we're not going to work with it at all in this course, but I just want you to see it, to see how this idea finally evolved. The thing we're trying to define is this. The lim as x goes to a of a f(x) is L. What does that mean to a modern mathematician? Here's what it means. Take a deep breath. For every $\varepsilon > 0$, there exists a $\delta > 0$ so that if $0 < |x - a|$, which in turn is $\times \delta$ then the $| f (x) - L | < \varepsilon$. That's a real mouthful there. The idea for every $\varepsilon > 0$, that's as close as one wishes and what you say is then there is a target value δ on the x axis. If x and a are within δ of each other, really close together than f (x) and L are within ε of each other you've hit the target, you're as close as one wishes.

As I say, we're not going to do anything more with this definition than look at it. It takes a lot of getting used to, but people who study mathematical analysis rather quickly do get used to it and from this we can in a perfectly rigorous fashion define all the concepts of calculus. Notice it has in it five inequalities if I count right, plus a universal statement, a for every statement, and an existential statement, there exists statement, and an implication in the bottom line. It's quite sophisticated, but such is the nature of modern calculus. That's my first big theme.

The second was liberation in which mathematics is freed from the bounds of reality. This came forth in the study of non-Euclidean geometry. Gauss remember had looked into this. He had studied a geometry, which he called non-Euclidean, in which triangles have fewer than 180 degrees. You have a triangle, you've got your angles α, β, γ add them up, it comes out less than 180 degrees. You say that's impossible. Triangles have 180 degrees in them. They do in Euclid's geometry, but if you look back when we were examining Euclid you see that the theorem that guaranteed that, as Euclid would've said, the sum of the angles of the triangle is two right angles. That hinged upon his parallel postulate. The parallel postulate is what Gauss jettisoned. Let's try a different postulate in that spot. All the other postulates were fine, but let's just replace the parallel postulate by an alternative that gives us triangles with less than 180 degrees in them and see what happens.

He's fiddling around with this and he said that you create in this fashion "… a special geometry, quite different from ours," by which he means Euclid's, but "which is absolutely consistent …" It's a perfectly good logical system.

Interestingly as Gauss was doing this, two other mathematicians were traveling the same path. It was one of these discoveries that were happening simultaneously in different places just as the calculus with Newton and Leibniz. This time the mathematicians were the Hungarian Johann Bolyai, who cooked up a geometry in which triangles have less than 180 degrees and the Russian Nikolai Lobachevski who cooked up a geometry in which triangles have less than 180 degrees. It was an amazing parallel, if I may use the pun there, where all three of these mathematicians were at work at the same time and we now attribute it to Gauss, Bolyai, and Lobachevski jointly. It sounds like a mathematical law firm. Gauss, Bolyai, and Lobachevski gave us non-Euclidean geometry.

It's an amazing realm. It's got surprising features. Bolyai said, "Out of nothing I have created a strange new universe." He's created this new geometry just with the power of his mind.

I want to show you one of the surprising results about this kind of geometry. It is one of these things that almost takes your breath away. How can this be? Yet if you look at the logic, you're bound to reach this conclusion. Follow me on this. First of all in the non-Euclidean realm a quadrilateral has less than 360 degrees in it I will argue. A quadrilateral, a four-sided figure, will have less than 360. That's pretty easy to see. If I take a quadrilateral ABCD, I want to show that the four angles add up to less than 360. Just draw a diagonal across it, draw AC, and that splits it into two triangles. Triangle ACD has less than 180 in it because this is non-Euclidean geometry, triangle ABC has less than 180 in it and if you add up the four angles A, B, C, and D—you see you're essentially adding up the six angles of the triangle and each of those triangles has less than 180, so the four angles of the quadrilateral have less than 180 + 180, 360. That's pretty clear. Quadrilaterals are deficient in their angles just as triangles are.

With that background, I claim this. In a non-Euclidean world, different size triangles have different angle sums! In Euclid's world every triangle has the same number of degrees in it, 180. That's not so in non-Euclidean geometry; different triangles have different sums. It's not like they all would have 179. They're different angle sums. There is no single value for the sum of the angles of a triangle.

Let me show you why. Here's a right triangle I will create in non-Euclidean geometry. Let's say that the horizontal and the vertical sides of this are both the same length. We'll call it x. That makes this an isosceles triangle and the theorems of isosceles triangles still hold in non-Euclidean geometry. That had nothing to do with parallels if you remember Euclid's development. This triangle has its base angles congruent. They're both α. There's this little triangle. Extend the horizontal leg out another x, extend the vertical leg upward another x, and connect the ends. Now you've got a great big triangle there. It's a right triangle, $2x$ on one side, $2x$ on the other. It's isosceles, so its base angles are the same. We'll call this β and β.

Then two other angles I want to fill in here are the angle next to the α if it's 180 degrees along a straight line and one piece is α, the other is 180 degrees minus α as is the one down below. I'm getting to something amazing here. Let me color in that quadrilateral that's the difference between the big triangle and the little triangle. Let me make it yellow. Remember from the yellow quadrilateral we can deduce that the sum of the four angles of the yellow quadrilateral is less than 360. If I have the $\beta + 180 - \alpha + 180 - \alpha + \beta$ as you go around, that comes out less than 360 degrees because all quadrilaterals have less than 360 degrees in the non-Euclidean world. But, look, there's a 180 on the left, another 180 on the left, they will cancel the 360 on the right, and when you collect your terms you see the $2\beta - 2\alpha$, which is what remains on the left is less than 0. In other words 2β is less than 2α so in the non-Euclidean geometry this would follow. Now look what happens. The sum of the angles of that great big triangle, the one we built on the outside is $90 + 2\beta$, but that's going to be less than $90 + 2\alpha$ because $2B < 2\alpha$, but that's the sum of the angles of the small triangle, $90 + 2\alpha$. What we just found is different size triangles have different angle sums. The angle sum of the big triangle is less than the angle sum of the little one. There is no single number that measures the angle sum of a triangle. That's strange. That's what Bolyai called the "strange new universe."

What mathematicians decided was if you studied Euclid's geometry where triangles all have 180 degrees, great. If you studied this non-Euclidean geometry where they don't great, but notice these both cannot be true of reality, maybe neither is, but certainly they can't both be true. When mathematicians are studying these two different branches at least one of them is wasting their time, at least in terms of reality. But, mathematicians say no, this is not a waste of time. Logically these are sound, these are consistent, these are legitimate pursuits, and thus mathematics is freed from the constraints of physical reality by the kinds of pursuits that Gauss, Bolyai, and Lobachevski began. It was a very major break.

That was pretty important. What about my third trend diversification, and there I mean diversification with regard to the cast of characters, the people who are allowed to play the game of mathematics? In particular, where are the women in this story? They faced insurmountable obstacles, almost insurmountable I guess, incredible obstacles to pursuing mathematics.

There was first of all a negative attitude. Women just weren't supposed to do this. This was not feminine and any woman seen doing mathematics was attacked. Second, there was the denial of a formal education. Mathematics is pretty sophisticated and as you move along you really need training. You need to go to school; you need to go to university to see what's happening here. If you're not allowed through the doors, you've got a great obstacle in your path. What about a job? Suppose you manage to overcome the negative attitude and to learn the math, what do you do with it? Is anyone going to hire a woman to do mathematics? These were very serious barriers.

In the 19th century, we start to see the door cracking open and here's one of the people that did this—Sophie Germain. A French mathematician, roughly lived at the time of Gauss, almost the same birthday. She broke through. She tells that as a youngster she'd read that story about Archimedes death. He's proving his theorem; he was so engrossed in this when the Roman came up, he told the Roman to go away and he gets killed. She thought, wow, if somebody could be that passionate about mathematics this is something I want to learn. Interestingly, that story inspired her.

She was self-taught in geometry and higher mathematics. She had to just learn this on her own, but she had the genius and she had the ability to push this through, to get a hold of things like Newton, Euler, Gauss, and start moving along in her mathematical training even though most of this had to be done in secret. She tells about her clandestine study by candlelight. Her parents wouldn't let her do this. After they went to bed, she'd sneak away and light a candle and do her mathematics like it was something illegal, but such was the negative attitude that this would've been reprimanded had she been caught.

Again this is just a horrible obstacle in your path, right. She didn't get to go to university; she was not allowed in. Women were not allowed in the classrooms back then, so sometimes she would eavesdrop at the door and try to hear what the mathematician's lectures were about or she had some sympathetic male colleagues that would share notes with her.

Eventually, she's up there right at the frontier doing real significant mathematics and she decided she wanted to write to Gauss. Germain writes to Gauss, but wasn't going to sign her name Sophie Germain, but rather she signed it M. LeBlanc. Gauss gets this letter in 1804, very impressive, from LeBlanc whoever this is. The correspondence continues for three years when Gauss discovers that his correspondent was a female, it was a woman, Sophie Germain. Gauss writes this. He says, "… when a woman, because of her sex, our customs, and prejudices … overcomes these fetters and penetrates that which is most hidden, she doubtless has the most noble courage, extraordinary talent, and superior genius." It was a very nice testimonial to what clearly was the courage of Sophie Germain to persist in the face of all these obstacles.

In 1816, she won the Paris Academy Prize for her work on elasticity, an area of applied mathematics. This was right in the spotlight—there she was, she won the prize. But, her best-known work today is in number theory where there is something called a Germain prime. We still study a number p is a Germain prime if $2p + 1$ is prime as well. This is an important distinction among the primes that she used as she was looking into Fermat's last theorem. If p is 5, for instance, that is a Germain prime because that's prime and $2p + 1$ is 11, which is also prime. Whereas $p = 7$ is a prime, it's not a Germain prime because $2(7) + 1$ is 15, which is composite. She left her marks in number theory.

Gauss wanted to honor her with an honorary doctorate from Göttingen, his university, and was in the process of doing so when she died in 1831. She didn't get that honor, but she made great contributions and started to open the door.

Other women followed in the 19th century, two I should mention. One is Sophia Kovalevskaya of Russia, a very interesting woman. She arranged a marriage to get out of Russia so she could go to Germany to study mathematics. She did, she worked with Weierstrass, later becomes the first woman to receive a professor appointment at a European university. The other is Grace Chisholm Young, English, studied the foundations of calculus and is the first woman to receive a Ph.D. from a German university. The 19th century saw the door crack open. I would not argue that it is wide open even today. There are still negative attitudes out there, but it is so much better than it was before these heroines pushed it open.

In my next lecture we're going to talk about that fourth great trend of the 19th century, abstraction, as it showed up in the set theory of Georg Cantor.

Cantor and the Infinite
Lecture 23

I find it impossible not to compare Cantor and Van Gogh. ... They lived at roughly the same time. ... They were both extremely innovative. ... They were both undeniably geniuses, but they both faced criticism for their radical work and they both suffered their mental demons.

As we saw in the last lecture, the 19th century was a period of rigorization, liberation, and diversification in mathematics. It was also a time of abstraction, a trend that is most evident in the work of Georg Cantor and his abstract set theory.

Cantor's work rests on two foundational principles, the first of which is the "completed" infinite, as opposed to the "potential" infinite. The idea of the potential infinite is that no matter where you are, you can always go further. With the natural numbers, for example, we can start with 1, 2, 3 and go further to 4, 5, 6, and so on. Cantor said that we can legitimately discuss the completed infinite—we can treat the natural numbers as a complete set.

The idea of the completed infinite can present significant paradoxes.

The idea of the completed infinite can present significant paradoxes. Consider, for instance, the tennis ball game between Jeff and Mutt. We give Jeff two numbered tennis balls and tell him to toss away the one with the higher number. We give Mutt two numbered tennis balls and tell him to toss away the one with the lower number. As we continue the game, we see that after each round, Jeff and Mutt seem to be left with a number of tennis balls equal to the number of rounds that have been played. If we could complete this infinite game, Jeff would have infinitely many odd-numbered tennis balls. We might think that Mutt would also have infinitely many tennis balls, but instead, he's left with none because at each step of the game, he has thrown away the lowest-numbered ball. This is a somewhat uncomfortable conclusion.

According to Cantor, the completed infinite is a fixed constant quantity lying beyond all finite magnitudes. The other premise on which Cantor built his theory of the infinite was a definition of the equal cardinality of two sets: Two sets have the same cardinality (that is, contain the same number of items) if their members can be put into a one-to-one correspondence with each other. Cantor proposed to apply this definition to sets that aren't necessarily finite, such as the set of natural numbers (N) and the set of even numbers. It seems that the two sets wouldn't match up in a one-to-one correspondence, but we can pair every n in the natural numbers with $2n$ in the even numbers: 1 with 2, 2 with 4,

Cantor explored the parodox of a completed infinite.

3 with 6, and so on. We can also find a one-to-one correspondence between the set of all integers (Z) and N. In Cantor's terminology, a set that can be put into a one-to-one correspondence with N is "denumerable."

We can define the number 4 by saying that a set has four members if it can be put into a one-to-one correspondence with the faces on Mt. Rushmore. Under this definition, there were four Beatles. What if we want to define a number that is infinite? Cantor introduced the number aleph-naught ($\aleph_0$) as the number that can be put into a one-to-one correspondence with the natural numbers. It is a transfinite cardinal number—perfectly legitimate but beyond the finite numbers.

This new idea leaves us with two big questions: First, can every infinite set be put into a one-to-one correspondence with N? Second, is there any mathematical use for this exploration of infinity? We'll see Cantor's answers in the next lecture. ■

Suggested Reading

Cantor, *Contributions to the Founding of the Theory of Transfinite Numbers.*

Dauben, *Georg Cantor.*

Grattan-Guinness, *From the Calculus to Set Theory.*

Russell, *The Autobiography of Bertrand Russell.*

Questions to Consider

1. Devise a modification of the tennis ball game with Jeff and Mutt and a third player—say, Emily—in which, after a completed infinitude of plays, Jeff is left with infinitely many tennis balls, Mutt with none, and Emily with exactly five. Does this sort of thing seem problematic, or are you comfortable (as was Cantor) with such outcomes?

2. Exhibit a formula that establishes a one-to-one correspondence between N, the set of all natural numbers, and O, the set of all odd natural numbers. Thus, your formula should match every whole number n with one and only one odd number and vice versa. In the terminology of Lecture 23, this shows that the set of odd whole numbers is denumerable.

Cantor and the Infinite
Lecture 23—Transcript

We're talking about the 19th century, the century of rigorization, liberation, and diversification. It is also the century of abstraction in mathematics. That is most evident in the work of Georg Cantor and his abstract set theory. I want to talk about it in this lecture and in the next. It's quite amazing.

This is Cantor. He was born in 1845 and lived into the 20th century. He came from a long line of musicians. People often make something of this, that he was little more artistic, that he was a little more of a romantic in his mathematics. You'll judge for yourself when you see it, but he was coming from a musical family. He studied with Weierstrass in Berlin. He was influenced by Weierstrass. Cantor's early work in number theory gave little hint of the radical direction that his mathematics would eventually take.

Cantor suffered from bouts of depression on a regular basis. He was hospitalized with his depression in 1884, 1899, 1902, 4, 7, and 11. He would go into the hospital, recover, come back, do some work, but then the depression would strike again, and he'd be back to the institution. He said that these periods cost him his mental freshness. We now can see this was a bipolar problem. There are treatments today; in those days, they would just institutionalize him for awhile.

He had some kind of quirky ideas outside of mathematics. He believed that Francis Bacon had written the works of Shakespeare and would speak and write about this. He was quite convinced that Bacon was the real author of Shakespeare's plays. He asserted that his mathematical discoveries had religious significance, that, in fact, God was leading him to certain mathematical achievements. If you're going to put forth some ideas that are truly radical in mathematics and you have these other issues—you know, touting the works of Shakespeare as being Bacon's products or telling people God was whispering over your shoulder, people can attack you. It's not just being eccentric; it was a source of ridicule. So, Cantor had to overcome such personal baggage as well as the intellectual challenge as evident in his mathematics. He ends up dying in a neuropathic hospital in 1918.

Unfortunately, he was institutionalized at the end and this was in World War I in Germany. It must've been a horrible, horrible ending.

I find it impossible not to compare Cantor and Van Gogh. I think this is a natural comparison. They lived at roughly the same time. They kind of resembled each other a little bit if you look at the pictures. They were both extremely innovative. They were doing things in their respective disciplines, be mathematics or painting, that hadn't been done before, pushing in new directions. They were both undeniably geniuses, but they both faced criticism for their radical work and they both suffered their mental demons. Cantor went back and forth to the mental hospital, Van Gogh put in a mental institution as well, and ultimately a suicide. There are certainly parallels in their lives. The one great difference is that Van Gogh died not knowing how he would be received, not knowing how greatly he would be loved by future generations. Cantor lived long enough to see his work actually prevail and so he had that satisfaction. For poor Vincent, he died having sold but one painting and probably figuring his life had been a failure. You would like to be able to tell him that now his works hang on museum walls around the world, reproductions are in homes all over the place. He is the great artist as we now recognize.

Cantor's great achievement, this radical thing that he did, was to tackle the infinite, to take on the infinite in a way that nobody ever had. His work rests on two foundational principles. Let me talk about the two principles that Cantor built upon. The first was that we can legitimately talk about the "completed" infinite, the completed infinite. What's that? We need a little background here.

No one doubted the legitimacy of the potential infinite. This was the dichotomy, the potential infinite versus the completed infinite. The potential infinite meant that no matter where you were you could always go further. If you have an infinite collection, you never get to the end. Here think of the natural numbers N and I have a little set bracket and then I have 1, 2, 3, if you're that far can you go further. Sure you can go to 4. If you go further sure it's a 5 to 6. You can keep going, the potential infinite is there, there's never going to be the last item. That's clear.

But, Cantor said we legitimately talk about completed infinite, we can complete that set, which I will indicate by writing down N, set bracket, 1, 2, 3, 4, 5, 6, … and I'll put the other bracket on the end. That is complete. Imagine that you have a bag and you're putting in the natural numbers 1, 2, 3; you're sticking them in the bag. Those who advocated the potential infinite would agree that you can always put more numbers in the bag. Cantor says we can complete the process; the bag is full. In there are all the natural numbers; it's a complete set. I can think about it, I can talk about it; I can treat it as a legitimate player.

In taking this position Cantor was running up against some important opposition, Gauss earlier in the century had rejected this. "I protest above all against the use of an infinite quantity as a completed one, which is never allowed," said Gauss. "The infinite is only a manner of speaking…" Gauss believed in the potential infinite, he most certainly did not believe in the completed one and so Cantor if he's going to come out in favor of the completed infinite he's taking on a pretty important opponent here, the great Gauss. It's hard for any mathematician to take on Gauss, Cantor does.

The completed infinite holds hidden dangers. It's true there are things going on with this concept that are troubling and I want to show you my favorite example of this, how if you grant me the completed infinite I can appear to run into some awful paradox. Watch this and see what you think.

This is going to be a multi-step game we're going to play, so I'll keep track of the steps, and I have two players. One of them we'll call Jeff. Here's what we do. In the first step I'm going to give Jeff two tennis balls, blue, two blue tennis balls with the numbers 1 and 2 on them. Jeff's job is to toss away the one with the higher number. He's got tennis balls 1 and 2 and he gives back tennis ball 2. Now he has a companion, a co-player, whom we will call Mutt. Mutt likewise gets two tennis balls on the first step, his are green, but his procedure is to toss away the one with the smaller number. Mutt gets one and two he throws away 1. At the end of the first step, how many tennis balls does each player have? Jeff has one tennis ball, number 1; Mutt has one tennis ball, number 2. They both have one. That's simple.

Here's step two. Give them each the next two tennis balls in order. Jeff will get tennis balls 3 and 4, blue. Mutt will get tennis balls 3 and 4, green. Jeff's job is to toss away the bigger number, so there goes the 4. Mutt's job is to toss away the smaller so at the moment Mutt has 2, 3, and 4 so he tosses the 2. After the second step how many does each have? Jeff has 1 and 3, Mutt has 3 and 4; they both have two tennis balls after two steps. Keep going to step three. We give Jeff tennis balls 5 and 6. We give Mutt tennis balls 5 and 6 of a different color. Now they play the game. Jeff throws away 6; Mutt throws away 3. After the third step, they each have three tennis balls. I'll just do one more; you see the pattern. In step four, Jeff gets 7 and 8, and Mutt gets 7 and 8. Jeff tosses the 8; Mutt tosses the 4. At this point, they both have four tennis balls.

If you continue this to seven steps, you can see each is going to have seven tennis balls and you can even see what they are. Jeff's tennis balls are going to be the first seven odd numbers because he always throws away an even. He'll have 1, 3, 5, 7, 9, 11, and 13, that's his 7. Mutt will have the 7 tennis balls from number 8 to number 14, but they both have 7 tennis balls after 7 steps. After 50 steps, they'd both have 50 tennis balls. After a million steps, they both have a million tennis balls.

Pretend that we can complete this process. The game is over. We have the completed infinite. We allow the process to be done. Question: How many tennis balls will each player have at the end of this completed process? Let's see here. That's pretty easy. Jeff's going to have infinitely many because he's going to have all the odd numbered ones. He has 1, and then 1 and 3, and then 1 and 3 and 5. He's going to have an infinitude of tennis balls—no problem.

It's Mutt that's the question. How many does he have left? Remember, after a billion steps, they both had a billion. After 50 billion steps, they both had 50 billion tennis balls. After infinitely many steps, think about it a minute, Mutt has none, no tennis balls. They're all gone. You say, wait a minute, that can't be. How can that be? Think about it. Mutt had the tennis balls 1, 2, 3, 4, 5, 6, there; they were given to him two at a time. On the first step, he threw away tennis ball 1, so he doesn't have that one left. At the second step, he threw away tennis ball 2, at the third step 3, at the fourth step 4, and then 5, and 6, and 7, and 8. He throws them all away. He can have nothing left at the end of this process. You say no, surely he has something left, maybe tennis ball number 1,000,000. No, he threw away tennis ball number 1,000,000 on the 1 millionth step. He throws them all away. If you grant me the completed infinite in this process I get what appears to be a kind of shocking conclusion that Mutt is out of tennis balls somehow. A lot of people find this very uncomfortable.

There's another way to think of this little tennis ball game though that might reveal what's going on a little more clearly. Suppose instead of getting these tennis balls two at a time, they just all got the tennis balls at once. They just got the infinitude at once, the completed infinite there they are. You can think what Jeff was doing is giving back the even ones, one at a time, so you give him infinitely many, he gives you back infinitely many, but if he gives you back the evens, he's still left with infinitely many whereas Mutt is giving them back one at a time, but he's giving them all back 1, 2, 3, 4 and he's got none left. You can see infinitely many in, infinitely many out could live you with infinitely many as in Jeff's case or infinitely many in, infinitely many out could leave you with none as in Mutt's case and you might be able to construct a game where infinitely many in, infinitely many out leaves you with 117 if you want to play it like that. The whole point though is that if you allow the completed infinite you'd better get ready for things like this. There're sort of hidden dangers and mathematicians that weren't ready for this were yelling about the completed infinite.

Gauss, remember, protested against the use of the infinite quantity as a completed one. Cantor responded and he said the completed infinite was a fixed constant quantity lying beyond all finite magnitudes. He believed in the completed infinite. That was one of his introductory premises to build his theory of the infinite. The other was this definition. He asked when did two sets have the same cardinality, by which he means the same number of items, when do two sets contain the same number of items, and a good example would be the sets of figures on my two hands. Do they have the same cardinality? Do I have the same number of fingers on each hand? If you ask anybody today they'd just say sure 1, 2, 3, 4, 5; 1, 2, 3, 4, 5, there's five fingers on each hand. That's fine, but Cantor said, no, we can answer this question of equal cardinality without having to count to five.

Suppose you couldn't count to five; I could still answer the question. Do I have the same number of fingers on each hand? The answer is you answer it by seeing if there is a 1:1 matching of the fingers on your two hands so that every finger on the left hand matches something on the right and vice versa and the matching is easily done, there it is. That's a 1:1 correspondence and that would show me that I have the same number of fingers on each hand even if I don't know the number five. His general definition is two sets have the same cardinality if their members can be put into a one-to-one correspondence with each other. Just think of matching fingers. If that can be done, yes, same cardinality, same number of items. If it can't be done then no, not the same cardinality.

You could imagine a problem like this. Suppose you're in a big auditorium, the crowd is coming in, there're lots of seats, lots of audience members, and somebody says are there as many seats as audience members. You say gee do I have to count all the seats? That'll take a long time and then count all the audience members, that'll take a very long time and then you could answer the question. Maybe there are 1,500 of each or something. Here's a much easier way to answer this question. Are there the same number of seats as audience members? You just get on the microphone and say would everybody please sit down. When that is accomplished if every person has a seat and every seat has a person then there's the same number of seats as audience members because that process of sitting down forms the 1:1 correspondence between people and seats. The answer would be yes in that case.

On the other hand, if when you tell everyone to sit down there's some people still standing then they don't have the same number of people and seats or if there're some empty seats over there the cardinalities are different. This actually seems like a fairly innocent concept right. It's certainly innocent with the fingers on your hand and with the people in the auditorium, but these are both finite sets—finitely many fingers, finitely many seats.

Cantor said why can't we apply this same definition of equal cardinality to sets that aren't necessarily finite? We can still ask the question if two infinite sets have the same number of elements by applying the very same definition, can you match up their members in a 1:1 fashion. This is pretty bold. He's out there. Are we going to talk about whether two infinite sets have the same cardinality? Here's an example of how this would work.

Let me let N be the natural numbers again, 1, 2, 3, 4, 5, 6, our old friend. I've got the set brackets on either side, suggesting that's a completed infinite. There is the completed infinite set, all the natural numbers are in there. Let me now introduce the even numbers, 2, 4, 6, 8, again in set brackets, that's that completed infinite. The question is do those two infinite sets have the same cardinality? Do they have the same number of items in them? It doesn't seem like it, right. It seems like there's only half as many even numbers as whole numbers and the answer should be no, but the answer is yes. They do have the same cardinality by Cantor's definition. What's he got to do? He's got to find a 1:1 correspondence matching every natural number with an even and every even with a natural and if he can do that, if he can find that correspondence that's like the fingers matched up left and right. That would be enough to guarantee they have equal cardinality.

Here comes the matching. Let's match 1 with 2 from the naturals to the evens, 2 with 4, 3 with 6, 4 with 8, 5 with 10 and so on. It's a natural way to do this; in general what I'm doing is taking N in the natural numbers and matching it with $2n$ in the evens. If you think about this, this is a 1:1 correspondence. Every natural number matches with some even, namely its double. Every even number matches with some natural number, namely its half. It's a perfectly good 1:1 correspondence. Conclusion is these two sets have the same cardinality by Cantor's definition and for him it was no different than the fingers on the hand or the people in the auditorium except these were infinite sets. It doesn't matter, he said; the same definition applies. There's just as many whole numbers as even numbers.

Let me go the other way. Let's look at the set Z for all the integers. We usually let Z stand for the set of all the integers positive, negative, and zero. It's the set as I've written it there in sort of doubly infinite—there's dot, dot, dot and you get –3, –2, –1, 0, 1, 2, 3, dot, dot, dot the other way. There's an infinite set. Can I find a 1:1 correspondence between Z and N, between the set of all integers and the set of natural numbers? Again, it might not seem like it. It seems like there's more integers than natural numbers because the natural numbers are already in there, but now you've got all these negatives we're throwing in. Doesn't that give you a bigger infinity? No, it doesn't. We're going to look for a 1:1 correspondence between N and Z.

Here's N, the set 1, 2, 3, 4, 5, and so on, and Z the set running from the negative through the negative integers through zero and out the other side to the positive ones. Now I'm going to try to find a 1:1 correspondence here. What you might try first would be let's take 1 from the natural numbers and match it with 1, 2 match with 2, 3 match with 3, 4 match with 4, and so on. There's a matching, but that's not going to work because if you think about it every natural number matches some integer, but not every integer gets matched with some natural number. There's nothing in this process that matches –2 or –3. That is not a 1:1 correspondence between all of N and all of Z. That fails. It didn't work.

But, just because my first try doesn't work doesn't mean there isn't another way to do this that will work, that will provide a 1:1 correspondence. In fact here it comes. I will match these two sets up in a 1:1 fashion. I'll start by matching 1 to 0 and here just to keep track of this let me make a little chart here. Here's N and its mate in Z so 1 is matched to 0. I'll match 2 to 1, so 2 matches with 1 from the natural numbers. Now it looks like I'm going down the wrong road here if I just keep going to the right, but now I'm going to match 3 with –1. I'll pick up the negative, so 3 matches with –1. Now 4 matches with 2, but then 5 goes and matches with –2, and so we continue this process matching the next positive, the next negative, the next positive, the next negative. If you want a formula for this you can actually crank one out here; the number N in the natural numbers matches this thing $1 + (-1)^n(2n - 1)/4$ in the integers. If you check this out, this is the correspondence I'm illustrating here. It is a 1:1 correspondence between N and Z. Conclusion is those two sets have the same cardinality.

Cantor says I'm going to introduce an adjective. I'm going to say that a set that can be put into a one-to-one correspondence with N, with the natural numbers, is called denumerable. Sometimes you'll see the term countably infinite used, but I'll use denumerable. If you can match it up with a set of natural numbers, we're going to say it is a denumerable set. That's an adjective. The set E is a denumerable set. The set Z is a denumerable set. We've seen in both cases the matching exists and of course N itself is a denumerable set because you can match N in a 1:1 fashion with itself just by matching 1 to 1, 2 to 2, 3 to 3, and so on.

But, now Cantor is ready for an even bolder move. He wants to introduce a new number, an infinitely large number. Maybe for this we better back up and just think about a finite case. Suppose you wanted to introduce the number 4. Suppose you wanted to tell people what 4 is. What does it mean for a set to have four things in it? You might never have thought about this. How do you define 4? Here comes a definition. I'm going to pose this one. I say a set has 4 members if it can be put into a 1:1 correspondence with the faces on Mt. Rushmore. That's going to be my standard of comparison, Mt. Rushmore, so if you want to see if you have four items in your set you've got to go see if you can match up your items with the heads on Mt. Rushmore. There's Mt. Rushmore and suppose somebody tosses me the set a, b, c, and d, and the question is does this set contain four members? It's up to me now to do a matching, so let's see here. We can match a with Jefferson, b with Roosevelt, c with Washington, and d with Lincoln. Every element in my set a, b, c, d has a president it's matched with. Every president has a letter it's matched with. By my definition, yes, there are four elements in my set. I've compared it to the standard, so Mt. Rushmore is my standard. How about this question? Are there four Beatles? Here are the Beatles; does it work? Sure, you just need a 1:1 correspondence between Beatles and presidents. We match Paul with Teddy, Ringo with Abe, John with Tom, and of course George with George. There's a 1:1 correspondence, so yes there are four Beatles.

Now Cantor said let's play the same game, but introduce a number that's infinite. He says I'm going to define a set to have aleph-naught members. Aleph-naught, we'll talk about that symbol in a minute, but here's the definition. If it can be put into a 1:1 correspondence with the natural numbers N, 1, 2, 3, 4, and so on, then we're going to tell how many members that set has. It has aleph-naught. Aleph-naught, aleph is the first letter of the Hebrew alphabet and the little sub-zero we would probably say alephs of zero, but the British version aleph-naught is what everybody says. Cantor chose this symbol because he wanted something different, a Hebrew letter as opposed to all the Latin and Greek letters that we're kicking around in mathematics because this really was different. He had just defined aleph-naught, which is a transfinite cardinal number, he said. It's a number, it's a perfectly legitimate number, but it's beyond the finite ones. It's transfinite, pretty exciting stuff here.

If I ask you how many Beatles are there, you would say four because you compared them to Mt. Rushmore, my definition of four. If I say how many even numbers are there, you'd say there's aleph-naught even numbers because of that 1:1 correspondence we showed between the evens and N. Aleph-naught is just as legitimate a number as 4—how many integers are there, aleph-naught, the same thing. We showed that correspondence. Now we have on the table this new kind of number, a transfinite quantity; nobody had ever gone down this road before. Cantor said when confronted about these kinds of controversial things, "The essence of mathematics lies in its freedom." He is free to think of the completed infinite. He is free to define sets with equal cardinality via the 1:1 correspondence and then he follows this where it leads and if it leads him into unknown territory so be it. It's a very bold direction for him to take.

At this point, we're left with two big questions at least. Here's the two I want to look at. One is can every infinite set be put into a one-to-one correspondence with N? The evens could; the integers could. Is this true of every infinite set? If the answer is yes, then there's sort of this one standard for infinity. You just match up the elements with the counting numbers you've got your infinite set. In other words, every infinite set would be denumerable by that definition. If the answer is no, then there's an infinite set that's so vast that you cannot match it with N. It would be—here's the scary thing—a bigger infinity. This is a pretty important question. Second question is, is any of this useful? Is there any mathematical use for this or is this just sort of a logical game? That's also a legitimate question and Cantor was confronted with this—so what, who cares?

In a remarkable paper from the year 1874 Cantor answered both of these questions. He answered them both in one paper. The answer to the first question is no. Can every infinite set be put into a one-to-one correspondence with N? No, there are bigger infinities and he showed where to find one. The answer to the second question, is this at all useful mathematically?

That answer is yes and Cantor shows a brilliant application of this in that 1874 paper.

My next lecture is going to be about those two questions. I'll answer them for you as Cantor did. Can every infinite set be put into a one-to-one correspondence with N and why does anybody care? The next lecture will show the answers.

Beyond the Infinite
Lecture 24

What Cantor has done here is found an infinite set that's bigger than N. N is the natural numbers—lots of infinite sets can be matched with it, but the set of all real numbers between 0 and 1 is so vast that it cannot be matched with N.

We ended the last lecture with two questions: Can every infinite set be matched in a one-to-one fashion with the natural numbers, and what good is any of this abstract set theory to other realms of mathematics?

To answer the first question, consider the fact that a real number can be written as an infinite decimal. For instance, 1/3 can be written as 0.33333...; π can be written as 3.141592.... We will regard the real numbers as the infinite decimals. A real number between 0 and 1 can be written as an infinite decimal, but its integer part will be 0, for example, $0.ABCDEF...$.

Let I (interval) be the set of all real numbers between 0 and 1. Cantor proved, through contradiction, that there cannot exist a one-to-one correspondence between N (the set of natural numbers) and I. If we assume the opposite, that such a correspondence exists, we might match up the natural numbers and the real numbers between 0 and 1 as follows: 1 with 0.123123123..., 2 with 0.9950000..., 3 with 0.602319888..., and so on. Let's now introduce a real number c of the form $0.c_1c_2c_3...$ and assume that the first decimal digit of c, c_1, is different from the first decimal digit of the number matched with 1. We choose 2 for c_1 and write c as $0.2c_2c_3...$. Next, c_2 must be different from the second decimal digit of the number matched with 2. Again, we choose any number other than 9 and continue the process. The number we generate belongs to I, but we have built it in such a way that it cannot be matched up with the numbers we have already matched to the natural numbers. We have reached a contradiction; thus, the set I of all real numbers between 0 and 1 is not denumerable. This is an infinite set that is larger than the infinite set N.

Cantor found an application for this work in the idea of an algebraic real number: A real number is algebraic if it is the solution to a polynomial with integer coefficients. By this definition, 2/3 qualifies as an algebraic number because it is the solution to the polynomial equation $3x - 2 = 0$. In fact, there are lots of algebraic numbers, some more complicated than others. Euler introduced the idea that there might be some real numbers that are not algebraic, that is, that could not be the solutions to any polynomial equation with integer coefficients. He called these non-algebraic numbers "transcendental" because they transcend algebra. Up until 1874, only two examples of transcendental numbers had been found; they seemed to be rare and difficult to establish. Then, Cantor wrote a paper in which he showed that the true situation is exactly the opposite: The transcendental numbers are by far more plentiful than the algebraic.

A landmark theorem has much in common with a landmark painting or a landmark novel. It possesses the aesthetic qualities of elegance and unexpectedness.

Cantor stands at the end of a long line of great thinkers that we have looked at in this course. In the first lecture, we noted that a landmark theorem has much in common with a landmark painting or a landmark novel. It possesses the aesthetic qualities of elegance and unexpectedness. I hope this course has indeed revealed to you the artistry of mathematics. ∎

Suggested Reading

Cantor, *Contributions to the Founding of the Theory of Transfinite Numbers*.

Dauben, *Georg Cantor*.

Dunham, *The Calculus Gallery*.

Grattan-Guinness, *From the Calculus to Set Theory*.

Russell, *The Autobiography of Bertrand Russell*.

1. Show that the number $1+\sqrt{5}$ is an algebraic number by exhibiting a specific polynomial equation with integer coefficients for which this number is a solution. This is done not by solving an equation but by "unsolving" one, as follows:

Let $x = 1+\sqrt{5}$. Then, $x-1 = \sqrt{5}$ and, thus, $(x-1)^2 = \left(\sqrt{5}\right)^2$. Expand and simplify this to get a quadratic equation with integer coefficients having $1+\sqrt{5}$ as a solution. This establishes that $1+\sqrt{5}$ is algebraic. NOTE: If you liked this exercise, apply similar reasoning to show that $\sqrt[5]{1+2\sqrt[3]{4}}$ is the solution to a (15th-degree) polynomial equation with integer coefficients and, thus, is also an algebraic number. This number was mentioned in Lecture 24.

2. The set of all fractions—be they positive, negative, or zero—is called the set of *rational* numbers. Cantor proved that this is a denumerable set. Use this and a cardinality argument to explain why there must exist real numbers that are not rational. These, of course, are known as irrationals. In your proof, how many irrationals did you explicitly exhibit?

3. By course's end, did you have a favorite great thinker? A favorite great theorem?

Beyond the Infinite
Lecture 24—Transcript

We've met Georg Cantor and his remarkable theory of the infinite, but now we have to address those two remaining questions. Can every infinite set be matched in a one to one fashion with the natural numbers and what good is any of this to other realms of mathematics?

To answer the first question, we're going to remember that a real number can be written as an infinite decimal. For instance, 1/3 we can write as 0.33333 forever. It's an infinite vessel. Two-fifths is 0.4 and you might just stop there. But, if we wanted, we could continue it as an infinite decimal with lots of zeros thereafter. Pi is 3.141592 and on it goes. There's a number like R, this curious pattern of 0.12123123412345—an infinite decimal. That's a real number. The real numbers we will regard as the infinite decimals.

A real number between 0 and 1, which is really where we're going to restrict our attention, can be written as an infinite decimal. But, its integer part is going to be 0 so it's going to look like 0 point and then here come the decimal places, A, B, C, D, E, F, and so on. But, if it lives between 0 and 1, that integer part to the left of the decimal point is just the 0.

Let I be the set of all the real numbers between 0 and 1. Here we go with one of these completed infinite sets. We'll call it I for interval. Here's the theorem Cantor proves. There cannot exist a 1:1 correspondence between N and I, between the natural numbers and the set of all real numbers between 0 and 1. There's no way you can make a 1:1 correspondence between these. Impossible! It's a pretty bold statement. How in the world do you prove this? How do you prove that the set of real numbers between 0 and 1 is not denumerable, in Cantor's terminology from the last lecture. The proof is by that age old weapon, contradiction. We've seen this in the hands of Archimedes. Here it is in modern times. Cantor's going to prove this by contradiction.

What's he trying to prove? There cannot be a 1:1 correspondence between N and I. How do you contradict it? You assume there is one. You assume there is such a 1:1 correspondence and get yourself into logical hot water. Let's assume, for the sake of eventual contradiction, that some 1:1 correspondence exists. What I have to do is display this hypothetical 1:1 correspondence. So let me say it's going to look something like this. It doesn't have to be this one, but I just need something on the screen, so let's say this is the correspondence. It could be anyone, the contradiction will be the same, but let's use this as my concrete examples.

What am I going to do? On the one hand, in the left-hand column the natural numbers N, so there's 1, 2, 3, 4, on they go. These are matched in a 1:1 fashion with real numbers between 0 and 1, which are these infinite decimals, so 1 has to match with something. Suppose it matches with the number 0.123123123. Suppose 2 matches with the number 0.9950000, 3 matches with 0.602319888; it could be anything and it matches with something—4 let's say matches with 0.7777 and so on. Hypothetically here it is. The 1:1 correspondence where in the left-hand column we have all the natural numbers. In the right-hand column we have all the real numbers between 0 and 1 and these are matched like the fingers on my hands. There's actually a logical contradiction lurking in front of us if we can just ferret it out.

Cantor says let's introduce a real number, which I'm going to call c and it's going to be a real number of the form $0.c_1c_2c_3$. There are the digits and my job is to figure out what these digits should be to get me into a contradiction. Let's suppose the first decimal digit of c, that is that thing that I'm calling c_1, is chosen to be different from the first decimal digit of the number matched with 1. That sounds a little confusing here, so let's go back to the correspondence. What I do is I look at the number matched with 1, so it's that 0.123123, I look at its first decimal digit, there it is put a circle around it. It's a 1. I want my c_1, my first digit of c not to be that. I want to insist that c_1 be something other than 1. Let's just crank it up 1 and make it 2, so c_1 is 2. Remember my number I'm trying to build here c looks like $0.c_1c_2c_3$. What I've just said is let c_1 be 2, so my number now looks like 0.2 and then here come c_2c_3 and the rest.

Cantor says choose c_2 to differ from the second decimal digit of the number matched with 2. Again back to the chart. The number matched with 2 is 0.995000. I go to the second decimal digit there, which is that 9, and I want to choose c_2 to be something other than 9. There are lots of choices; let's just crank it up to 0. My number c, which looked like $0.c_1c_2c_3$, now it starts off 0.20 and then here come the rest c_3c_4 and we continue in this process.

Let me do two more steps. I'll choose c_3 to differ from the third decimal digit of the number matched with 3. There we go. The number matched with 3, there it is 0.60231, etc. I go over to the third digit of that 602; I want c_3 to be something other than 2, so make it be 3. Now my number's looking like 0.203 and then the next digits are coming. C_4 will be chosen to differ from the fourth decimal digit of the number matched with 4. Back to the chart, the number matched with 4 is 0.77777; you go over to the fourth one, which is a 7 of course. I'll pick an 8. My number c starts 0.2038. This process can be continued forever until you build your whole real number, your infinite decimal c and this is called Cantor's diagonalization proof. You can see why from the picture, these numbers are spinning down the diagonal of that array.

You're asking so what? First of all the number c, that's generated 0.2038, belongs to I. Remember I was all the real numbers between 0 and 1 so is this a real number. Yes, it's an infinite decimal; c is an infinite decimal. Is it between 0 and 1, yes it starts 0.0, so it is a real number between 0 and 1; therefore, under Cantor's hypothesis that we had a 1:1 correspondence, c has to appear somewhere in that right-hand column. Remember we assumed every natural number was on the left, every number in I was on the right and they were matched up. If c is a number in I, it's got to be on the right somewhere.

But, it can't be. It's not there, not this number c. You can see it because c differs in at least one decimal place with every number in that right-hand column. You built it so it can't possibly be there. Look it can't be the number matched with 1 because that starts at 0.1 and you started at 0.2, so c is not that one. It can't be the number matched with 2 because there's a 9 in the second decimal place of the number matched with 2, but I have a 0 in the second decimal place of mine. It can't be the number matched with 3, it can't be the number matched with 4, c cannot be the number matched with 400 or 4 million—c isn't in the right-hand column. He has built a number that can't be there. Yet the hypothesis was that every number is in that right-hand column. Every number between 0 and 1 has got to be there somewhere;

contradiction has been reached. We have reached a contradiction. Conclusion is how did this all start? It started when Cantor assumed there was a 1:1 correspondence. Conclusion is there cannot be such a correspondence; it cannot exist. The set I of all real numbers between 0 and 1 is not denumerable.

It's usually called non-denumerable; it can't be matched. It's amazing; this is the proof. What Cantor has done here is found an infinite set that's bigger than N. N is the natural numbers, lots of infinite sets can be matched with it, but the set of all real numbers between 0 and 1 is so vast that it cannot be matched with N. It does not have cardinality aleph-naught, which is the cardinality we assigned to all sets that can be put into a 1:1 correspondence with N. It's a bigger creature. Cantor's just found a bigger infinity.

If you aren't comfortable with his work with the numerable sets in that first infinity, now you're really getting nervous because here comes a bigger infinity onto the scene. In fact it's not just the real numbers between 0 and 1; he shows that a bigger infinity exists for all the real numbers if you take all the ones across the real line. Again it's bigger than aleph-naught, so we have non-denumerable sets as well as denumerable ones. People are getting a little nervous about this point.

What about the application? We know there're sets that are so infinite they can't be matched with N, but so what. Who cares? Is there any use to this? This was put to Cantor and he realized that maybe he needed to sell this a little bit so he found an amazing application. This goes back to a definition that had been kicking around for a century and more. It says this; a real number is called algebraic if it is the solution to a polynomial equation with integer coefficients. It's called an algebraic number. If you can get it as a solution to a polynomial equation regardless of the degree, but the coefficients of that equation must be integers.

How about an example? I say the number 2/3 qualifies as an algebraic number by this definition. Why is that? I have to show it's the solution to some polynomial equation with integer coefficients. That's easy, $3x - 2 = 0$. If you put in $x = 2/3$ it certainly works. That's a solution to that linear equation, first-degree equation. It's a polynomial of degree 1 and notice the coefficients in that polynomial are integers 3 and –2. That's all you need 2/3 becomes an algebraic number.

I say the $\sqrt{3}$ is an algebraic number. If I claim that I've got to show that the $\sqrt{3}$ is a solution to a polynomial equation with integer coefficients. Again, that's not too hard; $x^2 - 3 = 0$ will work. That's a polynomial of degree 2, the coefficients are integers 1 and –3 and the $\sqrt{3}$ certainly solves that if you stick in $x = \sqrt{3}$ then the $\sqrt{3}^2 - 3$ is 0 so the $\sqrt{3}$ is an algebraic number.

They get more complicated. This thing, the $\sqrt[5]{1} + 2(\sqrt[3]{4})$ is an algebraic number, I claim because it's the solution to this polynomial equation with integer coefficients. Get read $x^{15} - 3x^{10} + 3x^5 - 33 = 0$. It turns out my fifth root up there at the top is a solution to that, that's a fifteenth degree polynomial with integer coefficients, so that's good enough. That makes my number algebraic. There're lots and lots of algebraic numbers.

It was Euler that introduced this concept. He said this is what the algebraic numbers are, but he thought there might be some others. There might be some real numbers that were not algebraic, which means they could not be the solution of any polynomial equation with integer coefficients, no matter the degree. He called these supposed non-algebraic numbers. He called them transcendental numbers because they transcend algebra. They transcend the polynomials; they go beyond what would fall out of polynomial equations as solutions. He imagines that there are these numbers, but the question is are there any? You might say well sure there are, he just gave it a name, transcendental numbers isn't that enough to guarantee there's something there? Well, no, that's not a guarantee there are any just because you've defined them to be all the rest of the numbers, maybe there aren't any more.

If you don't like that example, think of the world of animals. Suppose I am looking at dolphins and I say a dolphin is algebraic if it lives in the water. That's my definition, a dolphin is algebraic if it lives in the water and a dolphin will be transcendental if it's not algebraic. That's exactly the parallel that Euler was using between the algebraic numbers and the rest, the transcendental numbers. I've got my algebraic dolphins and the rest. That's a perfectly good definition of a transcendental dolphin, but there aren't any. There are no transcendental dolphins; you've never seen a dolphin walking down the street or flying overhead. They all live in the water. Just because I defined this category doesn't guarantee there are any members.

Are there any transcendental numbers? People started looking, but this is a very hard problem to show a number's transcendental. You have to show it can't be the solution of any polynomial equation with integer coefficients no matter the degree, so it could be a 15,000th degree polynomial, it can't be the solution of any such polynomial. It took a century before anybody found even one of these. The mathematician was Joseph Liouville; here he is. In 1844, he found the first example of a transcendental number. It was a very sophisticated argument, but he did it. Now we know there's at least one of these.

People then decided let's look at some famous numbers and see if they're transcendental and one of the numbers was *e*. We've met that number, Euler's number *e*. People speculated that that would be a good candidate to be a transcendental number so let's try to prove that. It was a big challenge and it took a lot of work and effort by Charles Hermite who in 1873 proved that *e* is a transcendental number; *e* is not the solution of any polynomial equation with integer coefficients. It was a great triumph, one of the great proofs of 19th-century mathematics.

People patted Hermite on the back and said great job. Here's another candidate, π. I think π is transcendental; Hermite, why don't you prove that? Since π is another famous number, it would be great to show that was transcendental. Hermite responded no way. He said, "I do not dare to attempt to show transcendence of π. If other undertake it no one will be happier than I about their success, but believe me my dear friend, this cannot fail to cost them some effort." He had worked so hard on *e* that he said one transcendental was enough so he wasn't going to even try π. It turned out π is transcendental, but that took a little longer to prove.

There's the situation. Transcendental numbers exist, but they seem kind of rare, hard to find, hard to establish. Then here comes Georg Cantor with his theory of the infinite in his 1874 paper and he shows that the true situation is exactly the opposite; that the transcendental numbers are more plentiful than the algebraic even though we are having trouble finding them. They're in the majority and he does this with his brilliant piece of reasoning, which I want to show you.

He turned intuition on its head and this was not the first time, nor would it be the last. What Cantor did was through a nice sophisticated argument he proved that the algebraic numbers could be put into a 1:1 correspondence with N, the natural numbers. These algebraic numbers, these solutions of polynomial equations with integer coefficients, he classified them in such a way that he got a match-up between the algebraic numbers and N even as we got a match-up in the previous lecture between Z and N, between the integers and N, or between e and N, between the evens and N. He showed how to do it with the algebraic matched up with N.

The algebraic numbers were denumerable because they match with N and their cardinality was aleph-naught, that famous transfinite number that Cantor had introduced. Those are the algebraic. Wait a minute; I showed you earlier in this lecture that he proved that the interval of real numbers between 0 and 1 was not denumerable. Indeed all the real numbers taken as a whole are not denumerable. The algebraic numbers are denumerable and all the real numbers are not and maybe you can see where this is going because the algebraic cannot account for all the real numbers. A denumerable set versus a non-denumerable set there's got to be more things than in the real numbers than just the algebraic and what are they. They're the transcendentals; they're out there. In fact, because denumerable is a small infinity and the non-denumerability of the reals is a big infinity it turns out not only are the transcendental there, but they're there in overwhelming abundance. The transcendental numbers are by far in the majority.

How about this little schematic? Suppose we have our balance, on one side Cantor puts the algebraic; on the other side he puts the reals. He knows that the algebraic are denumerable and the reals are not. The picture looks like this. They're in a balance; there's more weight on the side of the real numbers. Where'd the weight come from? How come there's more real numbers? It must be the transcendental tipping the scale. They're out there; they're out there in overwhelming abundance. Most numbers are transcendental and yet you know people had found so few of them; it seemed like such a battle to even find a couple.

What Cantor is saying is no, they're in the majority. E.T. Bell, a 20th-century mathematician and a very fine writer, put it this way: "The algebraic numbers," said Bell, "are spotted over the plane like stars against a black sky... the dense blackness is the firmament of the transcendental." The algebraic are out there, but they're like the stars just here and there, but that background, the blackness behind, are the transcendental, which are in the overwhelming majority.

There is an application of his theory the infinite, but wait a minute this was a cardinality argument, but how many transcendental did Cantor actually exhibit in this proof? Which ones did he show you? He showed none. He didn't show you a single one. He knew they were there because of this cardinality reason, but he didn't provide a specific example and people sort of wanted to see some. Come on; if you tell me the transcendentals are there, you tell me they're in the majority, show me a few. Cantor said I don't have to; I know they're there, I've counted.

That didn't go down too well with some people. Imagine, let's go back to our dolphins. Remember we got our algebraic dolphins and we got all our dolphins. Suppose I use some kind of clever counting argument and find that the scale tips as shown. What's that mean? The algebraic dolphins cannot account for all the dolphins so they'd be transcendental dolphins, which you remember are the dolphins that don't live in the water. If somebody claimed that there are dolphins that don't live in the water, I would sure want to see one. Show me one. But Cantor would say I don't have to. They're there; I know they're there. I counted. You can see how this could generate a bit of controversy, but it's a very nice piece of reasoning.

Was it just trickery? No, Cantor believed that this was not just playing games with infinity. This was a legitimate proof that led to a profound conclusion. He wrote: "I entertain no doubts as to the truths of the transfinites, which I recognized with God's help." He is certainly a fan of this fabulous theory that he invented.

Not everyone was so excited about it. Henri Poincaré, a 20th-century mathematician of great ability, said, "Later generations will regard set theory as a disease from which one has recovered." He thought this was sick, pathological. We don't really want to deal with this, but Poincaré actually was in the minority on this. Cantor ends up prevailing and we have a quotation from Bertrand Russell that says Cantor's work is "… probably the greatest of which the age can boast." That's pretty high praise from a very great mathematician and I think this is how the mathematical world today regards Cantor's contribution. It's indisputably great.

At this point I think we've reached the end of this lecture and we're reaching the end of the course. It seems like a fitting place to end to see Cantor using his wonderful imagination using the tools of logic like proof by contradiction and coming up with a result that is beautiful and really astonishing in its scope. It's exactly the sort of thing we've been trying to look at throughout this course. It places him in a line of great thinkers going way back.

We've seen a lot of them from Euclid and Archimedes and Heron in the time of the Greeks. Then we jumped over to medieval Islam and saw Al-Khwārizmī in action. We went back to Europe and the Renaissance with the crazy story of Cardano and the cubic. We encountered the mathematicians of the Heroic century, ending with Newton and Leibniz and the invention of calculus, and then their followers—the Bernoullis, Euler, later Gauss, and now Cantor. It's been a long string of great thinkers and a long string of great theorems that have accompany them.

I'd like to end with a few quotations, one from a very ancient mathematician from Classical times and one from our friend Bertrand Russell of the modern era. These obviously come from different time periods, but they both are getting at the power and beauty of mathematics. Proclus from the 5th century wrote this: "This, therefore, is mathematics: she gives life to her own discoveries; she awakens the mind and purifies the intellect; she brings light to our intrinsic ideas; she abolishes the oblivion and ignorance that are ours by birth." It's a pretty good testimonial to abolish oblivion and ignorance and give light to our ideas. Proclus captured something there about this wonderful discipline of mathematics.

Bertrand Russell also does this. We've seen him so many times. Here is a quotation I had in my very first lecture; I love it. He says mathematics is that "… ordered cosmos where pure thought can dwell as in its natural home."

There's this idea that mathematics is the realm where pure thought is most comfortable. Let me conclude with a quotation also from Russell that I think is a very apt one. He wrote this. He said:

> Mathematics, rightly viewed, possesses not only truth, but supreme beauty—a beauty cold and austere, like that of sculpture, without appeal to any part of our weaker nature, without the gorgeous trappings of painting or music, yet sublimely pure, and capable of a stern perfection such as only the greatest art can show.

Russell here has captured the essence of what we've been discussing throughout this course. Mathematics is great art. As I said at the outset, a landmark theorem has much in common with a landmark painting or a landmark novel. In the first lecture, I mentioned aesthetic qualities that a great proof can exhibit. One of these was elegance, an economy of thought in which the argument is reduced to its fundamental, rather like a haiku in which the author seems to say more than the words possibly can. Euclid's proof of the infinitude of primes is one of these. The elegance there was amazing or even Cantor's diagonalization proof we just looked at in which he established a very profound result in a very efficient economical elegant fashion.

Another characteristic of great theorems was their unexpectedness in which the strange nature of the result seems somehow to come out of the blue. Here I could cite Heron's formula for triangular area, very unexpected in my book. Euler's solution of the Basel Problem in which he summed that series to get $\pi^2/6$ or the proof we've just seen that the transcendental numbers vastly outnumber their seemingly more abundant algebraic brethren.

This course has been meant to reveal this sort of artistry in mathematics and to take us to that place where pure thought indeed can dwell as in its natural home. It's been a great pleasure to introduce you to these great thinkers and to share some of their masterworks. I hope you have enjoyed the journey.

Timeline

1850 B.C. .... Moscow Papyrus.

fl. c. 600 B.C. .... Thales.

c. 580–500 B.C. .... Pythagoras.

c. 395/390–342/337 B.C. .... Eudoxus, Greek mathematician who
introduced the method of exhaustion.

fl. c. 300 B.C. .... Euclid, author of *The Elements.*

287–212 B.C. .... Archimedes, author of the *Measurement
of a Circle* and other texts.

fl. c. 62... Heron of Alexandria.

c. 100–170 A.D. .... Ptolemy, author of *The Almagest.*

c. 500... Development of the numerals
in the base-10 system by
Indian mathematicians.

529.. Closing of the Library of Alexandria.

641.. Much of the materials of the Library of
Alexandria lost in a fire.

8th–13th centuries .... Golden age of Islamic mathematics.

c. 780–850...................................... Muhammad Mūsā ibn Al-Khwārizmī.

836–901... Thābit ibn Qurra, Islamic mathematician.

Timeline

1676.. Leibniz visits London and sees a manuscript of Newton's *De analysi.*

1684.. Leibniz publishes "*A Nova Methodus,*" "A New Method," the first-ever published paper on calculus.

1687.. Publication of Newton's *Principia Mathematica.*

1689.. The Basel problem issued by Jakob Bernoulli: find the exact sum of the infinite series $1 + 1/4 + 1/9 + 1/16 + 1/25…$; the problem was solved by Euler in the 1730s.

1696.. Publication of the first-ever textbook on calculus by the Marquis de l'Hospital, student of Johann Bernoulli; famous battle of the calculus wars initiated by Johann Bernoulli with his issuance of the brachistochrone problem.

1703.. Newton becomes president of the Royal Society.

1707.. Publication of Newton's textbook on algebra, the *Arithmetica Universalis,* which contained his proof of Heron's formula for the area of a triangle.

1707–1783...................................... Leonhard Euler.

1708.. The Royal Society, headed by Newton, attributes the development of calculus to Newton and obliquely accuses Leibniz of plagiarism.

1712.. The Royal Society, still headed by
Newton, restates its position attributing
the development of calculus to Newton.

1713.. Publication of the *Ars Conjectandi*, the
first treatise on probability theory, by
Jakob Bernoulli.

1736.. Posthumous publication of Newton's
treatise on "fluxions," i.e., calculus.

1740s .... Euler's work in the partitioning of
whole numbers.

1748.. Publication of Euler's *Introductio in
analysin infinitorum*, a textbook
on functions.

1750.. Euler finds 58 pairs of
amicable numbers.

1776–1831....................................... Sophie Germain, Self-taught
French mathematician.

1777–1855....................................... Carl Friedrich Gauss.

1789–1857....................................... Augustin-Louis Cauchy, French
mathematician credited with the
rigorization of calculus.

1793–1856....................................... Nikolai Lobachevski, Russian
mathematician who explored non-
Euclidean geometry.

1796.. Gauss's discovery of the regular 17-gon.

1799... Gauss proves the fundamental theorem of algebra in his doctoral dissertation.

1801... Publication of Gauss's *Disquisitiones arithmeticae*, a book on number theory.

1802–1860...................................... Johann Bolyai, Hungarian mathematician who explored non-Euclidean geometry.

1815–1897...................................... Karl Weierstrass, German mathematician who put forth the definition of limit.

1845–1918...................................... Georg Cantor.

1850–1891...................................... Sophia Kovalevskaya, Russian mathematician.

1868–1944...................................... Grace Chisholm Young, British mathematician and the first woman to receive a Ph.D. from a German university.

1874... Publication of Cantor's remarkable theory of the infinite.

1896... Discovery of Heron's work *Metrica* by R. Schöne in a library in Istanbul.

1911... Start of the project to publish Euler's complete works by the Swiss Academy of Science; the project is still ongoing.

Bibliography

Biggs, Norman, et al. *Graph Theory: 1736–1936*. New York: Clarendon Press, 1986. Euler's Königsberg bridge problem and all that followed.

Boyer, Carl. *The Concepts of the Calculus*. Wakefield, MA: Hafner Publishing, 1949. An older work and, thus, a bit dated, but a landmark discussion of the origins of calculus, with sections on Archimedes, Newton, Leibniz, and others.

———. *A History of Analytic Geometry*. New York: Scripta Mathematica, 1956. The tale of the algebra/geometry fusion, the greatest marriage in the history of mathematics.

Cantor, Georg. *Contributions to the Founding of the Theory of Transfinite Numbers*. New York: Dover (reprint), 1955. For those interested in a sophisticated look at Cantor's original writings on the transfinite; includes a long and useful introduction by Philip E. B. Jourdain.

Cardan, Jerome (Cardano, Gerolamo). *The Book of My Life* (translated by Jean Stoner). New York: Dover (reprint), 1962. The autobiography of the most colorful character from the history of mathematics. His story is a real hoot!

Cardano, Girolamo. *Ars Magna* (translated by T. Richard Witmer). New York: Dover (reprint), 1993. Cardano's 1545 textbook that contained, for the first time in print, the algebraic solution of the cubic equation in chapter XI. A mathematical classic.

Child, J. M., ed. *The Early Mathematical Manuscripts of Leibniz*. London: Open Court, 1920. Includes a translation of Leibniz's *Historia et Origo Calculi Differentialis*, his personal reminiscences about the invention of the calculus.

Dauben, Joseph. *Georg Cantor: His Mathematics and Philosophy of the Infinite*. Cambridge: Harvard University Press, 1979. Still the best introduction to the troubled genius who carried mathematics into the realm of the infinite.

Descartes, René. *The Geometry of René Descartes* (translated by D. E. Smith and Marcia Latham). New York: Dover (reprint), 1954. The first published analytic geometry in a treatise so opaque that it challenged both Newton and Leibniz. It is worth a look if only for that remarkable distinction.

Devlin, Keith. *The Unfinished Game*. New York: Basic Books, 2008. The story of the Pascal-Fermat correspondence that gave birth to the theory of probability.

Dunham, William. *The Calculus Gallery: Masterpieces from Newton to Lebesgue*. Princeton: Princeton University Press, 2005. A tour through an imagined museum of calculus, with rooms devoted to Newton, Leibniz, the Bernoullis, Euler, and others down to the 20th century. Mathematically sophisticated.

———. *Euler: The Master of Us All*. Washington, DC: Mathematical Association of America, 1999. A look at some of Euler's great theorems in subjects from number theory to infinite series to algebra. Lots of mathematics here.

———. *Journey through Genius: The Great Theorems of Mathematics*. New York: Wiley, 1990. More than any other, this book serves as the foundation of our course.

Dunnington, G. Waldo. *Carl Friedrich Gauss: Titan of Science*. Washington, DC: Mathematical Association of America, 2004. A nontechnical if somewhat dense biography of one of the greatest mathematicians of all.

Edwards, C. H. The *Historical Development of the Calculus*. New York: Springer-Verlag, 1979. This treatise, which pulls no mathematical punches, traces the development of the calculus from Archimedes to the 20th century.

Euclid. *Euclid's Elements*. Thomas L. Heath, trans., Dana Densmore, T.L. trans., Heath, Dana Densmore, ed. Santa Fe, NM: Green Lion Press, 2002. Anyone interested in the history of mathematics—or, for that matter, the history of Western civilization—should be acquainted with Euclid. This work, available here in a beautiful edition, provided the foundation for so much that followed.

Euler, Leonhard. *Elements of Algebra* (translated by John Hewlett). New York: Springer-Verlag, 1972. Here is Euler's great algebra text, a thorough and accessible masterpiece.

———. *Introduction to Analysis of the Infinite* (2 vols., translated by John Blanton). New York: Springer-Verlag, 1988, 1990. The English translation of Euler's classic *Introductio in analysin infinitorum*, which Carl Boyer called the "foremost textbook of modern times." Still an excellent read for the mathematically inclined.

Fauvel, John, et al. *Let Newton Be! A New Perspective on His Life and Works*. Oxford: Oxford University Press, 1988. A collection of articles about Newton's life and career, including sections on his mathematics, physics, and alchemy, as well as a discussion of his status as a British national hero.

Fellman, Emil. *Leonhard Euler*. Basel: Birkhäuser, 2007. A readable, informative, and "formula-free" biography of the great Euler.

Gjertsen, Derek. *The Newton Handbook*. London: Routledge and Kegan Paul, 1986. This curious volume contains everything anyone would want to know about Sir Isaac, arranged alphabetically by topic.

Grabiner, Judith. *The Origins of Cauchy's Rigorous Calculus*. New York: Dover (reprint), 2005. A serious mathematical examination of how Augustin-Louis Cauchy transformed the calculus with his introduction of "limits."

Grattan-Guinness, Ivor. *From the Calculus to Set Theory: 1630–1910*. Princeton: Princeton University Press, 1980. This splendid collection of essays covers the period from before Newton to after Cantor. Its mathematics is sophisticated; its overview of intellectual history is top-notch.

Hald, Anders. *A History of Probability and Statistics and Their Applications before 1750*. New York: Wiley, 1990. A scholarly look at Cardano, Fermat, Pascal, and Jakob Bernoulli—among others—as they established the theory of probability.

Hall, Rupert. *Philosophers at War*. Cambridge: Cambridge University Press, 1980. The Newton-Leibniz priority dispute in full detail.

Hardy, G. H. *A Mathematician's Apology*. Cambridge: Cambridge University Press, 1967. There is no more literate, engaging reflection about the beauty of mathematics than this, written at career's end by one of the 20th century's finest mathematicians.

Heath, Sir Thomas. *A History of Greek Mathematics*. New York: Dover (reprint), 1981. A survey of classical mathematics, featuring all the major players and a host of minor ones.

———. *The Works of Archimedes*. New York: Dover (reprint), 1953. This dense but wonderful volume contains the work of the greatest mathematician of ancient times, including the complete texts of *Measurement of a Circle* and *On the Sphere and the Cylinder*.

Heyne, Andreas, and Alice Heyne. *Leonhard Euler: A Man to Be Reckoned With*. Basel: Birkhäuser, 2007. The biography of Euler presented in pictures as a kind of "graphic novel." Believe it or not, it is quite enjoyable and, better yet, quite effective.

Hoffman, Joseph. *Leibniz in Paris: 1672–1676*. Cambridge: Cambridge University Press, 1974. An account of Leibniz's diplomatic mission to France when, in a few productive years, he advanced from mathematical novice to creator of the calculus.

Joseph, George Gheverghese. *The Crest of the Peacock: Non-European Roots of Mathematics*. New York: Penguin Books, 1991. A less technical, less comprehensive overview of non-Western mathematics than the Katz volume (below) but quite popular.

Katz, Victor, ed. *The Mathematics of Egypt, Mesopotamia, China, India, and Islam: A Sourcebook.* Princeton: Princeton University Press, 2007. An extraordinary volume, thorough in the extreme, that examines the broad sweep of non-Western mathematics.

Kline, Morris. *Mathematical Thought from Ancient to Modern Times.* Oxford: Oxford University Press, 1972. This remains our most comprehensive survey of the history of Western mathematics up to the 20th century.

Loomis, Elisha Scott. *The Pythagorean Proposition.* Washington, DC: National Council of Teachers of Mathematics, 1968. One of the quirkiest math books of all, a compendium of more than 400 different proofs of the Pythagorean theorem! Of particular interest to those who believe that variety is the spice of geometry.

Maor, Eli. *e: The Story of a Number.* Princeton: Princeton University Press, 1994. All anyone ever wanted to know about Euler's extraordinary number. The math is non-trivial.

Newton, Isaac. *The Correspondence of Isaac Newton* (7 vols.). Cambridge: Cambridge University Press, 1959–1977. Newton's letters, including correspondence with Hooke, Flamsteed, Halley, and the famous exchanges with Leibniz. This collection tends to be available only in research libraries, but it is worth a look if only to glimpse the man behind the legend.

Ore, Oystein. *Cardano: The Gambling Scholar.* New York: Dover (reprint), 1965. The biography of a true Renaissance man.

———. *Number Theory and Its History.* New York: McGraw-Hill, 1948. Although written in the mid-20th century, this remains an excellent introduction to number theory and its development from Euclid through Fermat, Euler, and Gauss. For good measure, the final chapter proves the constructability of the regular 17-gon .

Osen, Lynn. *Women in Mathematics.* Cambridge, MA: MIT Press, 1974. A nontechnical introduction to those women whose genius and courage broke gender barriers in the mathematical sciences, including chapters on Sophie Germain and Sofia Kovalevskaia.

Plofker, Kim. *Mathematics in India*. Princeton: Princeton University Press, 2009. A survey of the rich mathematical heritage of India.

Proclus. *A Commentary on the First Book of Euclid's Elements* (translated by Glenn Morrow). Princeton: Princeton University Press, 1970. A careful analysis of Euclid's Book I, from an ancient admirer.

Richeson, David. *Euler's Gem: The Polyhedron Formula and the Birth of Topology*. Princeton: Princeton University Press, 2008. This examines the famous formula $V + F = E + 2$ and many of its important consequences.

Robson, Eleanor. *Mathematics in Ancient Iraq: A Social History*. Princeton: Princeton University Press, 2008. A detailed look at the mathematics of ancient Mesopotamia.

————, and Jacqueline Stedall. *The Oxford Handbook of the History of Mathematics*. Oxford: Oxford University Press, 2009. A wonderful reference work detailing the sweep of mathematics from culture to culture and century to century.

Russell, Bertrand. *The Autobiography of Bertrand Russell, 1872–1914*. London: George Allen and Unwin Ltd., 1967. Covers the period of Russell's life when he was deeply engaged in the foundations of mathematics and exploring the consequences of Cantor's set theory. The writing gets full marks!

Sandifer, C. Edward. *How Euler Did It*. Washington, DC: Mathematical Association of America, 2007. A collection of essays describing some of Euler's most intriguing results. A great place to see mathematical genius in action.

Stedall, Jacqueline. *Mathematics Emerging: A Sourcebook, 1540–1900*. Oxford: Oxford University Press, 2008. A compendium of original sources that touch on many points covered in our course.

Struik, Dirk, ed. *A Source Book in Mathematics: 1200–1800*. Cambridge: Harvard University Press, 1969. Translations of mathematical masterpieces, ranging from number theory to algebra to calculus.

Swetz, Frank, and T. I. Kao. *Was Pythagoras Chinese?* University Park, PA: Penn State University Press, 1977. This little monograph with the provocative title describes Chinese discoveries about right triangles, all done independently of the Greeks.

Tent, M. B. W. *Leonhard Euler and the Bernoullis*. Natick, MA: A.K. Peters, 2009. A book for the young reader that outlines the lives of Euler and the Bernoulli clan.

———. *The Prince of Mathematics: Carl Friedrich Gauss*. Wellesley, MA: A.K. Peters, 2006. Like the previous entry, this book is aimed at younger readers. It surveys the contributions of one of the towering figures from the history of mathematics.

Trudeau, Richard. *The Non-Euclidean Revolution*. Boston: Birkhäuser, 1987. An introduction, with plenty of history, to the development of non-Euclidean geometry.

Weil, André. *Number Theory: An Approach through History*. Boston: Birkhäuser, 1984. A mathematically sophisticated examination of this fascinating subject, focusing especially on Fermat and Euler, written by one of the 20th century's great number theorists.

Westfall, Richard. *The Life of Isaac Newton*. Cambridge: Cambridge University Press, 1993. This abridgement includes the biographical information but omits the technical content of Westfall's more scholarly work, cited next.

———. *Never at Rest: A Biography of Isaac Newton*. Cambridge: Cambridge University Press, 1980. By far, the best "scientific biography" of Isaac Newton.

Whiteside, Derek, ed. *The Mathematical Works of Isaac Newton* (2 vols.). New York: Johnson Reprint Corporation, 1964, 1967. A sampler of Newton's mathematical writings, including his derivation of the sine series (vol. 1) and his proof of Heron's formula (vol. 2) .

Notes

Notes

Notes

Notes

Notes